A First Course in Probability:

Solutions Manual.

William J. Stewart

Department of Computer Science
North Carolina State University
Raleigh, NC 27695–8206
USA

Library of Congress Cataloging-in-Publication Data
Stewart, William J., 1946–
A First Course in Probability: Solutions Manual/William J. Stewart. -1st ed.

ISBN-10: 1499508743
ISBN-13: 978-1499508741

William J. Stewart
Professor of Computer Science
North Carolina State University
Raleigh, NC 27695, USA.

billy@ncsu.edu

1 3 5 7 9 10 8 6 4 2

Cover photograph: "Light and Shade", by William J. Stewart, 2013
Fundação Iberê Camargo, Porte Alegre, Brazil.

Contents

Chapter 1

The Essence of Probability

Exercise 1.2.1 A multiprocessing system contains six processors, each of which may be up and running, or down and in need of repair. Describe an element of the sample space and find the number of elements in the sample space. List the elements in the event $\mathcal{A} =$ "at least five processors are working."

Answer 1.2.1 Letting u signify that a processor is "up" and d signify that it is "down," an element of the sample space may be denoted by a string of six characters in which the ith character is u if the ith processor is "up" and is d if the ith processor is "down."

There are 2^6 elementary events in total, ranging from $uuuuuu$ (all six processors are up) through $dddddd$ (all six processors are down). The event $\mathcal{A} =$ "at least 5 processors are working" contains seven elements:

$$\mathcal{A} = \{uuuuuu,\ duuuuu,\ uduuuu,\ uuduuu,\ uuuduu,\ uuuudu,\ uuuuud\}$$

Exercise 1.2.2 A gum ball dispenser contains a large number of gum balls in three different colors, red, green, and yellow. Assuming that the gum balls are dispensed one at a time, describe an appropriate sample space for this scenario and list all possible events.

A determined child continues to buy gum balls until he gets a yellow one. Describe an appropriate sample space in this case.

Answer 1.2.2 Denoting the three possible colors as r, g and y for red, green, and yellow, respectively, an appropriate sample space is $\{r,\ g,\ y\}$. The set of all possible events is

$$\phi,\ \{r\},\ \{g\},\ \{y\},\ \{r,\ g\},\ \{r,\ y\},\ \{g,\ y\},\ \{r,\ g,\ y\}.$$

In the second scenario, an appropriate sample space is

$$\{y,\ \ ry, gy,\ \ rry, rgy, ggy,\ \ rrry, rrgy, rgry, rggy, grry, grgy, ggry, gggy, \ldots\}$$

Exercise 1.2.3 A brother and a sister arrive at the gum ball dispenser of the previous question, and each of them buys a single gum ball. The boy always allows his sister to go first. Let $\mathcal{A}$ be the event that the girl gets a yellow gum ball and let $\mathcal{B}$ be the event that at least one of them gets a yellow gum ball.

(a) Describe an appropriate sample space in this case.
(b) What outcomes constitute event $\mathcal{A}$?
(c) What outcomes constitute event $\mathcal{B}$?
(d) What outcomes constitute event $\mathcal{A} \cap \mathcal{B}$?
(e) What outcomes constitute event $\mathcal{A} \cap \mathcal{B}^c$?
(f) What outcomes constitute event $\mathcal{B} - \mathcal{A}$?

Answer 1.2.3

(a) Because the order is important in this example, we shall denote an element of the sample space by a pair, the first component of which gives the color of the gum ball that the girl receives and the second, the color of the one received by her brother. In this case, the sample space is given by

$$\{(r,r),\ (r,g),\ (r,y),\ (g,r),\ (g,g),\ (g,y),\ (y,r),\ (y,g),\ (y,y)\}$$

(b) $\mathcal{A} = \{(y,r),\ (y,g),\ (y,y)\}$, i.e., first component is yellow.

(c) $\mathcal{B} = \{(r,y),\ (g,y),\ (y,r),\ (y,g),\ (y,y)\}$, i.e., either, or both, components are yellow.

(d) $\mathcal{A} \cap \mathcal{B} = \{(y,r),\ (y,g),\ (y,y)\}$, i.e., elements that belong to both.

(e) $\mathcal{A} \cap \mathcal{B}^c = \phi$, i.e., elements in $\mathcal{A}$ that are not in $\mathcal{B}$.

(f) $\mathcal{B} - \mathcal{A} = \{(r,y),\ (g,y)\}$, i.e., elements in $\mathcal{B}$ that are not in $\mathcal{A}$.

Exercise 1.2.4 The mail that arrives at our house is for father, mother, or children and may be categorized into junk mail, bills, or personal letters. The family scrutinizes each piece of incoming mail and observes that it is one of nine types, from *jf* (junk mail for father) through *pc* (personal letter for children). Thus, in terms of trials and outcomes, each trial is an examination of a letter and each outcome is a two-letter word.

(a) What is the sample space of this experiment?
(b) Let $\mathcal{A}_1$ be the event "junk mail." What outcomes constitute event $\mathcal{A}_1$?
(c) Let $\mathcal{A}_2$ be the event "mail for children." What outcomes constitute event $\mathcal{A}_2$?
(d) Let $\mathcal{A}_3$ be the event "not personal." What outcomes constitute event $\mathcal{A}_3$?
(e) Let $\mathcal{A}_4$ be the event "mail for parents." What outcomes constitute event $\mathcal{A}_4$?
(f) Are events $\mathcal{A}_2$ and $\mathcal{A}_4$ mutually exclusive?
(g) Are events $\mathcal{A}_1$, $\mathcal{A}_2$ and $\mathcal{A}_3$ collectively exhaustive?
(h) Which events imply another?

Answer 1.2.4 Using the notation f, m, and c for father, mother, and children, respectively, and j, b, and p for junk, bills, and personal mail, respectively, we have

(a) The sample space is $\{jf,\ jm,\ jc,\ bf,\ bm,\ bc,\ pf,\ pm,\ pc\}$

(b) $\mathcal{A}_1 = \{jf,\ jm,\ jc\}$

(c) $\mathcal{A}_2 = \{jc,\ bc,\ pc\}$

(d) $\mathcal{A}_3 = \{jf,\ jm,\ jc,\ bf,\ bm,\ bc\}$

(e) $\mathcal{A}_4 = \{jf,\ jm,\ bf,\ bm,\ pf,\ pm\}$

(f) Events $\mathcal{A}_2$ and $\mathcal{A}_4$ are mutually exclusive since they have no elementary events in common.

(g) Events $\mathcal{A}_1$, $\mathcal{A}_2$, and $\mathcal{A}_3$ are not collectively exhaustive, since their union does not include pf nor pm.

(h) Event $\mathcal{A}_1$ implies $\mathcal{A}_3$ since $\mathcal{A}_1$ is a proper subset of $\mathcal{A}_3$. In words, if the mail is "junk mail" (event $\mathcal{A}_1$) then this implies that it is "not personal" (event $\mathcal{A}_3$).

Exercise 1.2.5 Consider an experiment in which three different coins (say a penny, a nickel and a dime in that order) are tossed and the sequence of heads and tails observed. For each of the following pairs of events, $\mathcal{A}$ and $\mathcal{B}$, give the subset of outcomes that defines the events and state whether the pair of events are mutually exclusive, collectively exhaustive, neither or both.

(a) $\mathcal{A}$: The penny comes up heads. $\mathcal{B}$: The penny comes up tails.
(b) $\mathcal{A}$: The penny comes up heads. $\mathcal{B}$: The dime comes up tails.
(c) $\mathcal{A}$: At least one of the coins shows heads.
$\mathcal{B}$: At least one of the coins shows tails.
(d) $\mathcal{A}$: There is exactly one head showing.
$\mathcal{B}$: There is exactly one tail showing.
(e) $\mathcal{A}$: Two or more heads occur. $\mathcal{B}$: Two or more tails occur.

Answer 1.2.5

(a) $\mathcal{A}$: The penny comes up heads. $\mathcal{B}$: The penny comes up tails.
$\mathcal{A} = \{HHH,\ HHT,\ HTH,\ HTT\}$; $\mathcal{B} = \{THH,\ THT,\ TTH,\ TTT\}$
The events $\mathcal{A}$ and $\mathcal{B}$ are mutually exclusive and collectively exhaustive.
They form a partition and constitute an event space.

(b) $\mathcal{A}$: The penny comes up heads. $\mathcal{B}$: The dime comes up tails.
$\mathcal{A} = \{HHH,\ HHT,\ HTH,\ HTT\}$; $\mathcal{B} = \{HHT,\ THT,\ HTT,\ TTT\}$
The events $\mathcal{A}$ and $\mathcal{B}$ are neither mutually exclusive (they have states HHT and HTT in common) nor collectively exhaustive (states THH and TTH are missing).

(c) $\mathcal{A}$: At least one of the coins shows heads. $\mathcal{B}$: At least one of the coins shows tails.
$\mathcal{A} = \{HHH,\ HHT,\ HTH,\ HTT,\ THH,\ THT,\ TTH\}$;
$\mathcal{B} = \{HHT,\ HTH,\ HTT,\ THH,\ THT,\ TTH,\ TTT\}$
The events $\mathcal{A}$ and $\mathcal{B}$ are not mutually exclusive but are collectively exhaustive.

(d) $\mathcal{A}$: There is exactly one head showing. $\mathcal{B}$: There is exactly one tail showing.
$\mathcal{A} = \{HTT,\ THT,\ TTH\}$; $\mathcal{B} = \{THH,\ HTH,\ HHT\}$
The events $\mathcal{A}$ and $\mathcal{B}$ are mutually exclusive but not collectively exhaustive.

(e) $\mathcal{A}$: Two or more heads occur. $\mathcal{B}$: Two or more tails occur.
$\mathcal{A} = \{HHH,\ HHT,\ HTH,\ THH\}$; $\mathcal{B} = \{TTT,\ TTH,\ THT,\ HTT\}$
The events $\mathcal{A}$ and $\mathcal{B}$ are mutually exclusive and collectively exhaustive.
They form a partition and constitute an event space.

Exercise 1.2.6 A brand new light bulb is placed in a socket and the time it takes until it burns out is measured. Describe an appropriate sample space for this experiment. Use mathematical set notation to describe the following events:

(a) $\mathcal{A} =$ the light bulb lasts at least 100 hours.
(b) $\mathcal{B} =$ the light bulb lasts between 120 and 160 hours.
(c) $\mathcal{C} =$ the light bulb lasts less than 200 hours.
(d) $\mathcal{A} \cap \mathcal{C}^c$

Answer 1.2.6 Since, theoretically, the light bulb can stop functioning at any instant of time, but will not last forever, an appropriate sample space is the set of nonnegative real numbers, $\Omega = \{t \mid 0 \leq t < \infty\}$.

(a) $\mathcal{A} = \{t \mid 100 \leq t < \infty\}$, i.e., the light bulb lasts at least 100 hours.

(b) $\mathcal{B} = \{t \mid 120 \leq t \leq 160\}$, i.e., the light bulb burns out between 120 and 160 hours.

(c) $\mathcal{C} = \{t \mid 0 \leq t < 200\}$, i.e., the battery lasts less than 200 hours.

(d) $\mathcal{A} \cap \mathcal{C}^c = \{t \mid 200 \leq t < \infty\}$, i.e., the light bulb lasts at least 200 hours.

Exercise 1.3.1 An unbiased die is thrown once. Compute the probability of the following events.

(a) $\mathcal{A}_1$: The number of spots shown is odd.
(b) $\mathcal{A}_2$: The number of spots shown is less than 3.
(c) $\mathcal{A}_3$: The number of spots shown is a prime number.

Answer 1.3.1 Denote the elementary events $1, 2, \ldots, 6$ corresponding to the number of spots that appear. Since the die is unbiased, the probability associated with each elementary event is $1/6$. Therefore

(a) $\text{Prob}\{\mathcal{A}_1\} = 3/6$, since $\mathcal{A}_1$ contains three elementary events, 1, 3, and 5.

(b) $\text{Prob}\{\mathcal{A}_2\} = 2/6$, since $\mathcal{A}_2$ contains two elementary events, 1 and 2.

(c) $\text{Prob}\{\mathcal{A}_3\} = 3/6$, since $\mathcal{A}_3$ contains three elementary events, 2, 3, and 5.

Exercise 1.3.2 Two unbiased dice are thrown simultaneously. Describe an appropriate sample space and specify the probability that should be assigned to each. Also, find the probability of the following events:

(a) $\mathcal{A}_1$: The number on each die is equal to 1.
(b) $\mathcal{A}_2$: The sum of the spots on the two dice is equal to 3.
(c) $\mathcal{A}_3$: The sum of the spots on the two dice is greater than 10.

Answer 1.3.2 An element of the sample space may be denoted by the pair (i, j), $i, j = 1, 2, \ldots, 6$, where i denotes the number of spots on the first die and j, the number on the second. This gives a total of 36 elementary events, all of which must have the same probability, 1/36.

(a) $\text{Prob}\{\mathcal{A}_1\} = 1/36$. Only one of the 36 elementary events, $(1, 1)$, has this outcome.

(b) $\text{Prob}\{\mathcal{A}_2\} = 2/36$. Only two elementary events, $(1, 2)$ and $(2, 1)$, give this result.

(c) $\text{Prob}\{\mathcal{A}_3\} = 3/36$. Only three elementary events, $(5, 6)$, $(6, 5)$ and $(6, 6)$, give this result.

Exercise 1.3.3 A card is drawn from a standard pack of 52 well-shuffled cards. What is the probability that it is a king? Without replacing this first king card, a second card is drawn. What is the probability that the second card pulled is a king? What is the probability that the first four cards drawn from a standard deck of 52 well-shuffled cards are all kings. Once drawn, a card is not replaced in the deck.

Answer 1.3.3 A standard deck of 52 cards contains four kings. Therefore, the probability of drawing one king is 4/52, or 1/13. This leaves 51 cards, of which three are kings. The probability of drawing one of these kings is 3/51. Continuing in this fashion, the probability of drawing four kings from a standard deck is

$$\frac{4}{52} \times \frac{3}{51} \times \frac{2}{50} \times \frac{1}{49} = \frac{24}{6,497,400} = 0.00000369.$$

Exercise 1.3.4 A card is drawn at random from a standard deck of 52 well-shuffled cards. Let $\mathcal{A}$ be the event that the card drawn is a queen and let $\mathcal{B}$ be the event that the card pulled is red. Find the probabilities of the following events and state in words what they represent.

(a) $\mathcal{A} \cap \mathcal{B}$.
(b) $\mathcal{A} \cup \mathcal{B}$.
(c) $\mathcal{B} - \mathcal{A}$.

Answer 1.3.4 The probability associated with events $\mathcal{A}$ and $\mathcal{B}$ are, respectively, given by

$$\text{Prob}\{\mathcal{A}\} = 4/52 \quad \text{and} \quad \text{Prob}\{\mathcal{B}\} = 26/52.$$

Then

$$\begin{aligned}
\text{(a) Prob}\{\mathcal{A} \cap \mathcal{B}\} &= 2/52, \\
\text{(b) Prob}\{\mathcal{A} \cup \mathcal{B}\} &= \text{Prob}\{\mathcal{A}\} + \text{Prob}\{\mathcal{B}\} - \text{Prob}\{\mathcal{A} \cap \mathcal{B}\} = 28/52, \\
\text{(c) Prob}\{\mathcal{B} - \mathcal{A}\} &= \text{Prob}\{\mathcal{B} \cap \mathcal{A}^c\} = \text{Prob}\{\mathcal{A} \cup \mathcal{B}\} - \text{Prob}\{\mathcal{A}\} \\
&= 28/52 - 4/52 = 24/52.
\end{aligned}$$

Part (a) gives the probability of drawing a red queen; part (b) gives the probability of pulling any red card or a queen and part (c), the probability of drawing any red card other than a queen.

Exercise 1.3.5 A university professor drives from his home in Cary to his university office in Raleigh each day. His car, which is rather old, fails to start one out of every eight times and he ends up taking his wife's car. Furthermore, the rate of growth of Cary is so high that traffic problems are common. The professor finds that 70% of the time, traffic is so bad that he is forced to drive past his preferred exit off the beltline, Western Boulevard, and take the next exit, Hillsborough street. What is the probability of seeing this professor driving to his office along Hillsborough street, in his wife's car?

Answer 1.3.5 Let A be the event "he uses his wife's car" and B be the event "he takes Hillsborough Street". We have

$$\text{Prob}\{A\} = \frac{1}{8}, \quad \text{Prob}\{B\} = \frac{7}{10}.$$

The desired answer is the intersection of these events. Since the reliability of his car is independent of traffic conditions, the answer is

$$\text{Prob}\{A \cap B\} = \text{Prob}\{A\} \times \text{Prob}\{B\} = \frac{1}{8} \times \frac{7}{10} = \frac{7}{80}.$$

Exercise 1.3.6 A prisoner in a Kafkaesque prison is put in the following situation. A regular deck of 52 cards is placed in front of him. He must choose cards one at a time to determine their color. Once chosen, the card is replaced in the deck and the deck is shuffled. If the prisoner happens to select three consecutive red cards, he is executed. If he happens to selects six cards before three consecutive red cards appear, he is granted freedom. What is the probability that the prisoner is executed.

Answer 1.3.6 The following events lead to the prisoner's execution:

1. The first three cards he draws are all red. This event occurs with probability 1/8.

2. The first card drawn is black, but the next three are red. This event occurs with probability 1/16.

3.
   - The first card is red, the second black and the next three red, or
   - The first card is black, the second black and the next three red.

   Both these events occur with probability 1/32, so their sum is 1/16.

4. The third card is black and the last three are red. The first two cards may be either black or red. This gives four possibilities each with probability 1/64. The sum in this case is 1/16.

Therefore, the probability that the prisoner is executed is the sum of the probabilities of these four events, which is equal to 5/16.

In terms of elementary events, we can take an elementary event to be any combination of red and black cards from $rrrrrr$ through $bbbbbb$, where r indicates a red card and b, a black card. Each event is equiprobable with probability 1/64. There are eight elementary events of the form $rrrxxx$, where x indicates a *don't care*; four elementary events of the form $brrrxx$; four elementary events of the form $xbrrrx$ and four of the form $xxbrrr$. Each of these elementary events lead to the prisoners execution. Since there are a total of 20 of them, the probability that the prisoner is executed is given by $20/64 = 5/16$.

Exercise 1.3.7 Three marksmen fire simultaneously and independently at a target. What is the probability of the target being hit at least once, given that marksman one hits a target nine times out of ten, marksman two hits a target eight times out of ten while marksman three only hits a target one out of every two times.

Answer 1.3.7 It is easier to find the probability that the target is missed by all three and to take the complement of this event. The probability that the target is missed by all three is the product of the probabilities that it is missed by each. (The intersection of these three events.) This is given by $.1 \times .2 \times .5 = .01$. The probability that the target is hit at least once, is therefore equal to 0.99.

Exercise 1.3.8 Fifty teams compete in a student programming competition. It has been observed that 60% of the teams use the programming language C while the others use C++, and experience has shown that teams who program in C are twice as likely to win as those who use C++. Furthermore, ten teams who use C++ include a graduate student, while only four of those who use C include a graduate student.

(a) What is the probability that the winning team programs in C?
(b) What is the probability that the winning team programs in C and includes a graduate student?
(c) What is the probability that the winning team includes a graduate student?
(d) Given that the winning team includes a graduate student, what is the probability that that team programmed in C?

Answer 1.3.8

(a) If the chances of winning were independent of the programming language, then since there are 30 teams using C, the probability that the winning team programmed in C would be $30/50 = 0.6$. However, teams who program in C are twice as likely to win as those who program in C++. To solve this problem, let p denote the probability that an individual team using C wins the competition and q the probability that an individual team programming in C++ wins. We implicitly assume that all teams programming in C are equally likely to win, and the same for teams programming in C++. This leads to the simultaneous equations

$$p = 2q, \quad 30p + 20q = 1,$$

which we can solve to obtain $p = 1/40$, $q = 1/80$. Since there are 30 teams who program in C and each has probability $1/40$ of winning, the probability that a team programming in C is the winner is $30 \times 1/40 = 0.75$.

(b) Only four of the 30 teams that program in C include a graduate student. Each of these 30 teams has probability of $1/40$ of winning. The answer is therefore $4/40 = 0.1$

(c) The probability that the winning team programs in C++ and includes a graduate student is $10/80$. The probability that the winning team includes a graduate student is therefore

$$\frac{4}{40} + \frac{10}{80} = 0.225.$$

(d) Consider only those teams that include a graduate student. There are 14 of them. We now have $4p + 10q = 1$ and $p = 2q$. It follows this time, that $p = 1/9$ and $q = 1/18$. The probability that the winning team programmed in C, given that the team included a graduate student is therefore $4/9$.

Exercise 1.4.1 Let $\mathcal{A}$ be the event that an odd number of spots comes up when a fair die is thrown, and let $\mathcal{B}$ be the event that the number of spots is a prime number. What is Prob$\{\mathcal{A}|\mathcal{B}\}$ and Prob$\{\mathcal{B}|\mathcal{A}\}$?

Answer 1.4.1 Event $\mathcal{A}$ occurs if a 1, 3 or 5 appears and event $\mathcal{B}$ occurs is a 2, 3 or 5 appears. Therefore

$$\begin{aligned} \text{Prob}\{\mathcal{A}|\mathcal{B}\} &= 2/3, \\ \text{Prob}\{\mathcal{B}|\mathcal{A}\} &= 2/3. \end{aligned}$$

Exercise 1.4.2 A card is drawn from a well-shuffled standard deck of 52 cards. Let $\mathcal{A}$ be the event that the chosen card is a heart, let $\mathcal{B}$ be the event that it is a black card, and let $\mathcal{C}$ be the event that the chosen card is a red queen. Find $\text{Prob}\{\mathcal{A}|\mathcal{C}\}$ and $\text{Prob}\{\mathcal{B}|\mathcal{C}\}$. Which of the events $\mathcal{A}$, $\mathcal{B}$, and $\mathcal{C}$ are mutually exclusive?

Answer 1.4.2 Event $\mathcal{A}$ contains the 13 hearts and has probability $1/4$, event $\mathcal{B}$ contains the 26 black cards and has probability $1/2$ and event C contains the two red queens and has probability $1/26$. Also, $\text{Prob}\{\mathcal{A}|\mathcal{C}\} = 1/2$ and $\text{Prob}\{\mathcal{B}|\mathcal{C}\} = 0$.

The events $\mathcal{A}$ and $\mathcal{B}$ are mutually exclusive as are the events $\mathcal{B}$ and $\mathcal{C}$, since they have no elementary events in common. Events $\mathcal{A}$ and $\mathcal{C}$ are not mutually exclusive.

Exercise 1.4.3 A family has three children. What is the probability that all three children are boys? What is the probability that there are two girls and one boy? Given that at least one of the three is a boy, what is the probability that all three children are boys. You should assume that $\text{Prob}\{boy\} = \text{Prob}\{girl\} = 1/2$.

Answer 1.4.3 This question is similar to that of tossing three fair coins and examining the number of heads obtained. In this case we take the elementary events to be a string of length 3, with either a b (boy) or g (girl) in each position. The sample space contains eight elementary events, from bbb through ggg. With this information, we compute the answers as follows:

The probability of three boys is $(1/2)^3 = 1/8$. The probability of two girls and one boy is $3/8$, since the boy can be the first second, or third child. To answer the final part, we observe that seven of the eight elementary events contain at least one boy, and only one of these seven have three boys. Therefore

$$\text{Prob}\{bbb \mid \text{at least 1 boy}\} = 1/7.$$

Alternatively, if $\mathcal{A}$ is the event bbb (three boys) and $\mathcal{B}$ is the event *at least one boy*, then event $\mathcal{A}$ contains a single outcome while event $\mathcal{B}$ contains seven including the outcome in $\mathcal{A}$. Thus $\text{Prob}\{\mathcal{A} \cap \mathcal{B}\} = 1/8$; $\text{Prob}\{\mathcal{B}\} = 7/8$ and

$$\text{Prob}\{\mathcal{A}|\mathcal{B}\} = \frac{\text{Prob}\{\mathcal{A} \cap \mathcal{B}\}}{\text{Prob}\{\mathcal{B}\}} = \frac{1/8}{7/8} = 1/7.$$

Exercise 1.4.4 Three cards are placed in a box; one is white on both sides, one is black on both sides, and the third is white on one side and black on the other. One card is chosen at random from the box and placed on a table. The (uppermost) face that shows is white. Explain why the probability that the hidden face is black is equal to $1/3$ and not $1/2$.

Answer 1.4.4 The card with two black sides has been eliminated and one of the remaining two cards shows a white face. The hidden face could be black or white. It may be thought that each of the two remaining cards has equal chance of being the one on the table and so the probability that the hidden face is black is one-half. But this is incorrect.

The problem should be addressed by observing that after the double black card has been eliminated, there remain four faces, three of which are white and the last one black. Each of these four faces has an equal chance of appearing as the top face on the table but also each has an equal chance of being the hidden face. Once the top face is exposed, this leaves three possibilities for the face on the bottom. If the top face is white, then the bottom face has two possibilities of being white and one of being black, all three being equally probable. Hence the correct answer of 1/3. In terms of conditional probabilities, we have

$$\text{Prob}\{\text{black on bottom}|\text{white on top}\}$$
$$= \frac{\text{Prob}\{\text{black on bottom \& white on top}\}}{\text{Prob}\{\text{ white on top}\}} = 1/3,$$

since there is only one possibility of having "black on bottom and white on top", while there are three possible ways to have "white on top".

Exercise 1.5.1 If $\text{Prob}\{\mathcal{A}\,|\,\mathcal{B}\} = \text{Prob}\{\mathcal{B}\} = \text{Prob}\{\mathcal{A}\cup\mathcal{B}\} = 1/2$, are $\mathcal{A}$ and $\mathcal{B}$ independent?

Answer 1.5.1 From the data given, we need to find $\text{Prob}\{\mathcal{A}\}$ and $\text{Prob}\{\mathcal{A}\cap\mathcal{B}\}$ in order to see if $\text{Prob}\{\mathcal{A}\cap\mathcal{B}\} = \text{Prob}\{\mathcal{A}\}\text{Prob}\{\mathcal{B}\}$.
We can find $\text{Prob}\{\mathcal{A}\cap\mathcal{B}\}$ from the relationship

$$1/2 = \text{Prob}\{\mathcal{A}\,|\,\mathcal{B}\} = \frac{\text{Prob}\{\mathcal{A}\cap\mathcal{B}\}}{\text{Prob}\{\mathcal{B}\}} = \frac{\text{Prob}\{\mathcal{A}\cap\mathcal{B}\}}{1/2},$$

i.e., $\text{Prob}\{\mathcal{A}\cap\mathcal{B}\} = 1/4$. We may now find $\text{Prob}\{\mathcal{A}\}$ from the relationship

$$\text{Prob}\{\mathcal{A}\cup\mathcal{B}\} = \text{Prob}\{\mathcal{A}\} + \text{Prob}\{\mathcal{B}\} - \text{Prob}\{\mathcal{A}\cap\mathcal{B}\},$$

i.e.,

$$\text{Prob}\{\mathcal{A}\} = \text{Prob}\{\mathcal{A}\cup\mathcal{B}\} - \text{Prob}\{\mathcal{B}\} + \text{Prob}\{\mathcal{A}\cap\mathcal{B}\} = 1/2 - 1/2 + 1/4 = 1/4.$$

We may now conclude that $\mathcal{A}$ and $\mathcal{B}$ are *not* independent since

$$1/4 = \text{Prob}\{\mathcal{A}\cap\mathcal{B}\} \neq \text{Prob}\{\mathcal{A}\}\text{Prob}\{\mathcal{B}\} = (1/4)(1/2) = 1/8.$$

Exercise 1.5.2 A flashlight contains two batteries that sit one on top of the other. These batteries come from different batches and may be assumed to be independent of one another. Both batteries must work in order for the flashlight to work. If the probability that the first battery is defective is 0.05 and the probability that the second is defective is 0.15, what is the probability that the flashlight works properly?

Answer 1.5.2 Let $\mathcal{A}_1$ be the event that the first battery is defective and let $\mathcal{A}_2$ be the event that the second battery is defective. Then

$$\text{Prob}\{\mathcal{A}_1\} = 0.05 \quad \text{and} \quad \text{Prob}\{\mathcal{A}_2\} = 0.15.$$

The probability that the flashlight does not work is $\text{Prob}\{\mathcal{A}_1 \cup \mathcal{A}_2\}$, since if either battery fails, the flashlight does not work. This probability may be found as

$$\text{Prob}\{\mathcal{A}_1 \cup \mathcal{A}_2\} = \text{Prob}\{\mathcal{A}_1\} + \text{Prob}\{\mathcal{A}_2\} - \text{Prob}\{\mathcal{A}_1 \cap \mathcal{A}_2\}.$$

Since $\mathcal{A}_1$ and $\mathcal{A}_1$ are independent, $\text{Prob}\{\mathcal{A}_1 \cap \mathcal{A}_2\} = \text{Prob}\{\mathcal{A}_1\}\text{Prob}\{\mathcal{A}_2\}$ and so

$$\text{Prob}\{\mathcal{A}_1 \cup \mathcal{A}_2\} = 0.05 + 0.15 - (0.05)(0.15) = 0.1925.$$

It follows then that the probability that the flashlight works is $1.0 - 0.1925 = 0.8075$.
Alternatively, the answer can be obtained as

$$(1 - \text{Prob}\{\mathcal{A}_1\}) \times (1 - \text{Prob}\{\mathcal{A}_2\}) = 0.95 \times 0.85 = 0.8075.$$

Exercise 1.5.3 A spelunker enters a cave with two flashlights, one that contains three batteries in series (one on top of the other) and another that contains two batteries in series. Assume that all batteries are independent and that each will work with probability 0.9. Find the probability that the spelunker will have some means of illumination during his expedition.

Answer 1.5.3 All three batteries in the first flashlight or else both batteries in the second flashlight must work in order for the caver to have light. Let $\mathcal{A}_i$, for $i = 1, 2, \ldots, 5$, denote the event that the ith battery works. Then, the probability that the first flashlight works is $\text{Prob}\{\mathcal{A}_1 \cap \mathcal{A}_2 \cap \mathcal{A}_3\}$, while for the second it is $\text{Prob}\{\mathcal{A}_4 \cap \mathcal{A}_5\}$ The probability that the caver has light is

$$\text{Prob}\{\text{Has light}\} = \text{Prob}\{(\mathcal{A}_1 \cap \mathcal{A}_2 \cap \mathcal{A}_3) \cup (\mathcal{A}_4 \cap \mathcal{A}_5)\}$$

$$= \text{Prob}\{\mathcal{A}_1 \cap \mathcal{A}_2 \cap \mathcal{A}_3\} + \text{Prob}\{\mathcal{A}_4 \cap \mathcal{A}_5\} - \text{Prob}\{\mathcal{A}_1 \cap \mathcal{A}_2 \cap \mathcal{A}_3 \cap \mathcal{A}_4 \cap \mathcal{A}_5\}.$$

Now, due to the independence of the five events, we have

$$\begin{aligned} \text{Prob}\{\mathcal{A}_1 \cap \mathcal{A}_2 \cap \mathcal{A}_3\} &= \text{Prob}\{\mathcal{A}_1\}\text{Prob}\{\mathcal{A}_2\}\text{Prob}\{\mathcal{A}_3\} = (0.9)^3, \\ \text{Prob}\{\mathcal{A}_4 \cap \mathcal{A}_5\} &= \text{Prob}\{\mathcal{A}_4\}\text{Prob}\{\mathcal{A}_5\} = (0.9)^2, \end{aligned}$$

$$
\begin{aligned}
&\text{Prob}\{\mathcal{A}_1 \cap \mathcal{A}_2 \cap \mathcal{A}_3 \cap \mathcal{A}_4 \cap \mathcal{A}_5\} \\
&= \text{Prob}\{\mathcal{A}_1\}\text{Prob}\{\mathcal{A}_2\}\text{Prob}\{\mathcal{A}_3\}\text{Prob}\{\mathcal{A}_4\}\text{Prob}\{\mathcal{A}_5\} = (0.9)^5,
\end{aligned}
$$

and

$$\text{Prob}\{\text{Has light}\} = (0.9)^3 + (0.9)^2 - (0.9)^5 = 0.9485.$$

Exercise 1.6.1 Six boxes contain white and black balls. Specifically, each box contains exactly one white ball; also box i contains i black balls, for $i = 1, 2, \ldots, 6$. A fair die is tossed and a ball is selected from the box whose number is given by the die. What is the probability that a white ball is selected?

Answer 1.6.1 The balls are partitioned into six mutually exclusive and collectively exhaustive sets. Each has probability $1/6$ of being chosen. Also, the probability of selecting the white ball from a set of one white and i black balls is $1/(1+i)$. Hence, from the theorem of total probability, we have

$$\text{Prob}\{\text{white}\} = \frac{1}{6}\left(\frac{1}{2} + \frac{1}{3} + \frac{1}{4} + \frac{1}{5} + \frac{1}{6} + \frac{1}{7}\right) = 0.2655.$$

Exercise 1.6.2 A card is chosen at random from a deck of 52 cards and inserted into a second deck of 52 well-shuffled cards. A card is now selected at random from this augmented deck of 53 cards. Show that the probability of this card being a queen is exactly the same as the probability of drawing a queen from the first deck of 52 cards.

Answer 1.6.2 There are four queens in the original deck of 52 cards so the probability of drawing one is $1/13$. To find the probability of selecting a queen from the augmented deck of 53 cards, let $\mathcal{A}$ be the event that a queen is moved from the first deck and inserted into the second. Then $\mathcal{A}^c$ is the event that some card other than a queen is selected and inserted into the second deck. Let $\mathcal{B}$ be the event that a queen is selected from the second deck. Then

$$
\begin{aligned}
\text{Prob}\{\mathcal{B}\} &= \text{Prob}\{\mathcal{B} \,|\, \mathcal{A}\}\text{Prob}\{\mathcal{A}\} + \text{Prob}\{\mathcal{B} \,|\, \mathcal{A}^c\}\text{Prob}\{\mathcal{A}^c\} \\
&= \frac{5}{53} \times \frac{1}{13} + \frac{4}{53} \times \frac{12}{13} = \frac{1}{13}.
\end{aligned}
$$

Exercise 1.6.3 A factory has three machines that manufacture widgets. The percentages of a total day's production manufactured by the machines are 10%, 35%, and 55%, respectively. Furthermore, it is known that 5%, 3%, and 1% of the outputs of the respective three machines are defective. What is the probability that a randomly selected widget at the end of the day's production runs will be defective?

Answer 1.6.3 Let $\mathcal{B}_i$, $i = 1, 2, 3$, be the event that the widget is produced by machine i, and let $\mathcal{A}$ be the event that the selected widget is defective. Then, from the law of total probability,

$$\text{Prob}\{\mathcal{A}\} =$$

$$\text{Prob}\{\mathcal{A}|\mathcal{B}_1\}\text{Prob}\{\mathcal{B}_1\} + \text{Prob}\{\mathcal{A}|\mathcal{B}_2\}\text{Prob}\{\mathcal{B}_2\} + \text{Prob}\{\mathcal{A}|\mathcal{B}_3\}\text{Prob}\{\mathcal{B}_3\}$$
$$= (0.05)(0.1) + (0.03)(0.35) + (0.01)(0.55) = 0.0210.$$

Exercise 1.6.4 A computer game requires a player to find safe haven in a secure location where her enemies cannot penetrate. Four doorways appear before the player, from which she must choose to enter one and only one. The player must then make a second choice from among two, four, one, or five potholes to descend, respectively depending on which door she walks through. In each case one pothole leads to the safe haven. The player is rushed into making a decision and in her haste makes choices randomly. What is the probability of her safely reaching the haven?

Answer 1.6.4 Let $\mathcal{A}$ be the event that the player gets to the safe haven. Let $\mathcal{B}_i$ be the event that she first elects doorway i. These are our mutually exclusive events, since the player must choose one of them. Since one is chosen at random, we have $\text{Prob}\{\mathcal{B}_i\} = 1/4$ for each i. For $i = 1$, we are told that there are now two additional choices (potholes) one of which leads to the safe haven. Thus $\text{Prob}\{\mathcal{A}|\mathcal{B}_1\} = 1/2$. Similarly

$$\text{Prob}\{\mathcal{A}|\mathcal{B}_2\} = 1/4, \quad \text{Prob}\{\mathcal{A}|\mathcal{B}_3\} = 1, \quad \text{Prob}\{\mathcal{A}|\mathcal{B}_4\} = 1/5.$$

The probability of safely arriving is now given from the theorem of total probability as

$$\begin{aligned}\text{Prob}\{\mathcal{A}\} &= \text{Prob}\{\mathcal{A}|\mathcal{B}_1\}\text{Prob}\{\mathcal{B}_1\} + \text{Prob}\{\mathcal{A}|\mathcal{B}_2\}\text{Prob}\{\mathcal{B}_2\} \\ &+ \text{Prob}\{\mathcal{A}|\mathcal{B}_3\}\text{Prob}\{\mathcal{B}_3\} + \text{Prob}\{\mathcal{A}|\mathcal{B}_4\}\text{Prob}\{\mathcal{B}_4\} \\ &= \frac{1}{4}\left(\frac{1}{2} + \frac{1}{4} + 1 + \frac{1}{5}\right) = 0.4875.\end{aligned}$$

Exercise 1.6.5 The first of two boxes contains b_1 blue balls and r_1 red balls; the second contains b_2 blue balls and r_2 red balls. One ball is randomly chosen from the first box and put into the second. When this has been accomplished, a ball is chosen at random from the second box and put into the first. A ball is now chosen from the first box. What is the probability that it is blue?

Answer 1.6.5 After the balls have been exchanged, there are three possibilities

1. $\mathcal{A}_1$: The number of blue balls in the first box has not changed.
2. $\mathcal{A}_2$: A blue ball in the first box has been replaced by a red ball

3. $\mathcal{A}_3$: A red ball in the first box has been replaced by a blue ball

The event $\mathcal{A}_1$ may be divided into two disjoint events: "The first is blue and the second is blue" and "The first is red and the second is red". Hence the probability of event $\mathcal{A}_1$ may be written as

$$\text{Prob}\{\mathcal{A}_1\} = \frac{b_1}{b_1+r_1} \times \frac{b_2+1}{b_2+r_2+1} + \frac{r_1}{b_1+r_1} \times \frac{r_2+1}{b_2+r_2+1}.$$

The probabilities of the other two events are simply given as

$$\text{Prob}\{\mathcal{A}_2\} = \frac{b_1}{b_1+r_1} \times \frac{r_2}{b_2+r_2+1},$$

$$\text{Prob}\{\mathcal{A}_3\} = \frac{r_1}{b_1+r_1} \times \frac{b_2}{b_2+r_2+1}.$$

Let $\mathcal{B}$ denote the event, "A blue ball is chosen from the first box". Then

$$\text{Prob}\{\mathcal{B}\} =$$

$$\text{Prob}\{\mathcal{B}|\mathcal{A}_1\}\text{Prob}\{\mathcal{A}_1\} + \text{Prob}\{\mathcal{B}|\mathcal{A}_2\}\text{Prob}\{\mathcal{A}_2\} + \text{Prob}\{\mathcal{B}|\mathcal{A}_3\}\text{Prob}\{\mathcal{A}_3\}$$

and we have

$$\text{Prob}\{\mathcal{B}|\mathcal{A}_1\} = \frac{b_1}{b_1+r_1},$$

$$\text{Prob}\{\mathcal{B}|\mathcal{A}_2\} = \frac{b_1-1}{b_1+r_1},$$

$$\text{Prob}\{\mathcal{B}|\mathcal{A}_3\} = \frac{b_1+1}{b_1+r_1}.$$

Taking this altogether gives

$$\text{Prob}\{\mathcal{B}\} = \frac{b_1}{b_1+r_1} + \frac{r_1b_2 - b_1r_2}{(b_1+r_1)^2(b_2+r_2+1)}.$$

Exercise 1.7.1 Returning to Exercise 1.6.1, given that the selected ball is white, what is the probability that it came from box 1?

Answer 1.7.1 Let $\mathcal{A}$ be the event that the chosen ball is white and let $\mathcal{B}_i$, $i = 1, 2, \ldots, 6$, be the event that the selected ball came from box i. We need to find $\text{Prob}\{\mathcal{B}_1 \,|\, \mathcal{A}\}$. From Question 1.6.1, we found that $\text{Prob}\{\mathcal{A}\} = 0.2655$. Also,

$$\text{Prob}\{\mathcal{A} \,|\, \mathcal{B}_1\} = 0.5 \quad \text{and} \quad \text{Prob}\{\mathcal{B}_1\} = 0.1667.$$

Hence

$$\text{Prob}\{\mathcal{B}_1 \,|\, \mathcal{A}\} = \frac{\text{Prob}\{\mathcal{A} \,|\, \mathcal{B}_1\}\text{Prob}\{\mathcal{B}_1\}}{\text{Prob}\{\mathcal{A}\}} = \frac{(0.5)(0.1667)}{0.2655} = 0.3139.$$

Exercise 1.7.2 In the scenario of Exercise 1.6.3, what is the probability that a defective, randomly selected widget was produced by the first machine? What is the probability that it was produced by the second machine. And the third?

Answer 1.7.2 Again, we let $\mathcal{B}_i$, $i = 1, 2, 3$ be the event that the defective widget is produced by machine i, and $\mathcal{A}$ be the event that the selected widget is defective. We apply Bayes law and seek $\text{Prob}\{\mathcal{B}_1 \,|\, \mathcal{A}\}$. We have

$$\text{Prob}\{\mathcal{B}_1 \,|\, \mathcal{A}\} = \frac{\text{Prob}\{\mathcal{A} \,|\, \mathcal{B}_1\}\text{Prob}\{\mathcal{B}_1\}}{\text{Prob}\{\mathcal{A}\}}.$$

The answer to Exercise 1.6.3 gives

$$\text{Prob}\{\mathcal{A}\} = 0.0210,$$

while the statement of the problem gives

$$\text{Prob}\{\mathcal{A} \,|\, \mathcal{B}_1\} = 0.05 \quad \text{and} \quad \text{Prob}\{\mathcal{B}_1\} = 0.1.$$

Therefore

$$\text{Prob}\{\mathcal{B}_1 \,|\, \mathcal{A}\} = \frac{(0.05)(0.1)}{0.0210} = 0.2381.$$

Similarly, we have

$$\text{Prob}\{\mathcal{B}_2 \,|\, \mathcal{A}\} = \frac{(0.03)(0.35)}{0.0210} = 0.5000$$

and

$$\text{Prob}\{\mathcal{B}_3 \,|\, \mathcal{A}\} = \frac{(0.01)(0.55)}{0.0210} = 0.2619.$$

Observe that the sum of all three is equal to 1.

Exercise 1.7.3 A bag contains two fair coins and one two-headed coin. One coin is randomly selected, tossed three times, and three heads are obtained. What is the probability that the chosen coin is the two-headed coin?

Answer 1.7.3 Let $\mathcal{A}$ be the event that three heads are obtained and let $\mathcal{B}$ be the event that the two-headed coin is selected. We require $\text{Prob}\{\mathcal{B} \,|\, \mathcal{A}\}$. We have

$$\text{Prob}\{\mathcal{B}\} = 1/3, \quad \text{Prob}\{\mathcal{B}^c\} = 2/3, \quad \text{Prob}\{\mathcal{A} \,|\, \mathcal{B}\} = 1$$
$$\text{and} \quad \text{Prob}\{\mathcal{A} \,|\, \mathcal{B}^c\} = 1/8.$$

Hence

$$\text{Prob}\{\mathcal{B} \,|\, \mathcal{A}\} = \frac{\text{Prob}\{\mathcal{A}|\mathcal{B}\}\text{Prob}\{\mathcal{B}\}}{\text{Prob}\{\mathcal{A}|\mathcal{B}\}\text{Prob}\{\mathcal{B}\} + \text{Prob}\{\mathcal{A}|\mathcal{B}^c\}\text{Prob}\{\mathcal{B}^c\}}$$
$$= \frac{1/3}{1/3 + 2/24} = \frac{4}{5}.$$

Exercise 1.7.4 A most unusual Irish pub serves only Guinness and Harp. The owner of this pub observes that 85% of his male customers drink Guinness as opposed to 35% of his female customers. On any given evening, this pub owner notes that there are three times as many males as females. What is the probability that the person sitting beside the fireplace drinking Guinness is female?

Answer 1.7.4 Let $\mathcal{A}$ be the event "drinks Guinness", and $\mathcal{A}^c$ be the event "drinks Harp". Let $\mathcal{F}$ be the event "the customer beside the fireplace is female, and $\mathcal{F}^c$ the event "the customer beside the fireplace is male". Then we have

$$\text{Prob}\{\mathcal{A}|\mathcal{F}\} = .35, \quad \text{Prob}\{\mathcal{A}^c|\mathcal{F}\} = .65, \quad \text{Prob}\{\mathcal{A}|\mathcal{F}^c\} = .85,$$

$$\text{Prob}\{\mathcal{A}^c|\mathcal{F}^c\} = .15, \quad \text{Prob}\{\mathcal{F}\} = .25, \quad \text{Prob}\{\mathcal{F}^c\} = .75.$$

We wish to calculate $\text{Prob}\{\mathcal{F}|\mathcal{A}\}$. From Bayes' law, we have

$$\begin{aligned}\text{Prob}\{\mathcal{F}|\mathcal{A}\} &= \frac{\text{Prob}\{\mathcal{A}|\mathcal{F}\}\text{Prob}\{\mathcal{F}\}}{\text{Prob}\{\mathcal{A}|\mathcal{F}\}\text{Prob}\{\mathcal{F}\} + \text{Prob}\{\mathcal{A}|\mathcal{F}^c\}\text{Prob}\{\mathcal{F}^c\}} \\ &= \frac{.35 \times .25}{.35 \times .25 + .85 \times .75} = .1207\end{aligned}$$

Exercise 1.7.5 Historically, on St. Patrick's day (March 17), the probability that it rains on the Dublin parade is 0.75. Two television stations are noted for their weather forecasting abilities. The first, which is correct nine times out of ten, says that it will rain on the upcoming parade; the second, which is correct eleven times out of twelve, says that it will not rain. What is the probability that it will rain on the upcoming St. Patrick's day parade?

Answer 1.7.5 Consider the following events:

- $\mathcal{A}$: It rains on St. Patrick's day.
- B: The first TV station says that it will rain on the parade while the second says that it will not.
- TV_1: The first TV station is correct.
- TV_2: The second TV station is correct.

We have

$$\text{Prob}\{\mathcal{A}\} = 0.75, \quad \text{Prob}\{\mathcal{A}^c\} = 0.25, \quad \text{Prob}\{TV_1\} = 0.9, \quad \text{Prob}\{TV_2\} = 11/12.$$

We wish to find the conditional probability $\text{Prob}\{\mathcal{A}|B\}$ which we shall compute from Bayes' Law. To use this law, we will need $\text{Prob}\{B|\mathcal{A}\}$ and $\text{Prob}\{B|\mathcal{A}^c\}$. We have

$$\text{Prob}\{B|\mathcal{A}\} = \text{Prob}\{TV_1\}\text{Prob}\{{TV_2}^c\} = \frac{9}{10} \times \frac{1}{12} = 0.075$$

and

$$\text{Prob}\{B|\mathcal{A}^c\} = \text{Prob}\{TV_1{}^c\}\text{Prob}\{TV_2\} = \frac{1}{10} \times \frac{11}{12} = 0.0917.$$

Now applying Bayes' Law we obtain

$$\text{Prob}\{\mathcal{A}|B\} = \frac{\text{Prob}\{B|\mathcal{A}\}\text{Prob}\{\mathcal{A}\}}{\text{Prob}\{B|\mathcal{A}\}\text{Prob}\{\mathcal{A}\} + \text{Prob}\{B|\mathcal{A}^c\}\text{Prob}\{\mathcal{A}^c\}} =$$

$$\frac{0.075 \times 0.75}{0.075 \times 0.75 + 0.0917 \times 0.25}, = 0.7105$$

which is the probability that it will rain on the upcoming parade.

Exercise 1.7.6 80% of the murders committed in a certain town are committed by men. A dead body with a single gunshot wound in the head has just been found. Two detectives examine the evidence. The first detective, who is right seven times out of ten, announces that the murderer was a male but the second detective, who is right three times out of four, says that the murder was committed by a woman. What is the probability that the author of the crime was a woman?

Answer 1.7.6 Consider the following events

- $\mathcal{M}$: The murderer is male.
- $\mathcal{D}$: The first detective says the murderer is male and the second says the murderer is female.
- $\mathcal{A}_1$: The first detective is correct.
- $\mathcal{A}_2$: The second detective is correct.

We have

$$\text{Prob}\{\mathcal{M}\} = 0.8, \quad \text{Prob}\{\mathcal{M}^c\} = 0.2, \quad \text{Prob}\{\mathcal{A}_1\} = 0.7, \quad \text{Prob}\{\mathcal{A}_2\} = 0.75.$$

The probability that the author of the crime is male is equal to the conditional probability $\text{Prob}\{\mathcal{M}|\mathcal{D}\}$. To compute this we first find

$$\text{Prob}\{\mathcal{D}|\mathcal{M}\} = \text{Prob}\{\mathcal{A}_1\} \times \text{Prob}\{\mathcal{A}_2^c\} = 0.7 \times .25 = .175$$

and

$$\text{Prob}\{\mathcal{D}|\mathcal{M}^c\} = \text{Prob}\{\mathcal{A}_1^c\} \times \text{Prob}\{\mathcal{A}_2\} = 0.3 \times .75 = .225.$$

We can now apply Bayes' law to obtain

$$\text{Prob}\{\mathcal{M}|\mathcal{D}\} = \frac{\text{Prob}\{\mathcal{D}|\mathcal{M}\}\text{Prob}\{\mathcal{M}\}}{\text{Prob}\{\mathcal{D}|\mathcal{M}\}\text{Prob}\{\mathcal{M}\} + \text{Prob}\{\mathcal{D}|\mathcal{M}^c\}\text{Prob}\{\mathcal{M}^c\}} =$$

$$\frac{0.175 \times 0.8}{0.175 \times 0.8 + 0.225 \times 0.2} = 0.7568.$$

The probability that the author of the crime is a woman is therefore 0.2432.

Exercise 1.8.1 Write code similar to that given in the text to simulate each of the three scenarios described below. Also, for each scenario, compute the exact result mathematically and compare your simulation answers with the exact answer.

(a) A prisoner in a Kafkaesque prison is put in the following situation. A regular deck of 52 cards is placed in front of him. He must choose cards one at a time to determine their color. Once chosen, the card is replaced in the deck and the deck is shuffled. If the prisoner happens to select three consecutive red cards, he is executed. If he happens to select six cards before three consecutive red cards appear he is granted freedom. What is the probability that the prisoner is executed?

(b) Three cards are placed in a box; one is white on both sides, one is black on both sides, and the third is white on one side and black on the other. One card is chosen at random from the box and placed on a table. The (uppermost) face that shows is white. What is the probability that the hidden face is black?

(c) A factory has three machines that manufacture widgets. The percentages of a total day's production manufactured by the machines are 10%, 35%, and 55% respectively. Furthermore, it is known that 5%, 3%, and 1% of the outputs of the respective three machines are defective. What is the probability that a randomly selected widget at the end of the day's production runs will be defective?

Answer 1.8.1 The Java code and the results obtained with a single run of this code are given for all three scenarios, along with the exact result.

(a)
```
// ***  Begin Simulation ***************************
        int dead= 0;
        for (int n=1; n<max[i]; n++) {   // Generate max[i] runs
            int ncards = 0;  int nreds = 0; int d = 0;
            while(ncards < 6 && d==0) {
                int r_e = generator.nextInt(2);    ncards++;
                if (r_e == 1) { nreds++;
                    if (nreds == 3) {d++;}
                }
                else {nreds = 0;}
            }
            if (d > 0){dead++;}
        }
        double prob = (double)dead/max[i]; //Approximate probability
// ***  End Simulation ***************************
```

n	50	100	1,000	10,000	50,000	100,000	500,000	1,000,000
Prob	0.22	0.33	0.303	0.3058	0.31282	0.31332	0.311874	0.312717

Table 1.1: Probability that the prisoner is executed.

The exact answer is $5/16 = 0.3125$.

(b)
```
// ***  Begin Simulation ***************************
        int n_e = 0;     int n_c = 0;
        for (int n=1; n<max[i]; n++) {      // Generate max[i] runs
            int r_e = generator.nextInt(3);
            if (r_e > 0){                   // WW or WB card pulled
                r_e = generator.nextInt(2);
                if (r_e > 0) {              // WB card pulled
                    r_e = generator.nextInt(2);
                    if (r_e == 0){n_c++; n_e++;}   // B on bottom
                }
                else {n_c++;}
            }
        }
         double prob = (double)n_e/n_c;    // Approximate probability
// ***  End Simulation ***************************
```

n	50	100	1,000	10,000	50,000
Prob	0.272727	0.290910	0.326489	0.325424	0.3308455

n	100,000	500,000	1,000,000
Prob	0.330895	0.332629	0.333655

Table 1.2: Probability that the bottom card is black.

The exact answer is 1/3.

(c)
```
// ***  Begin Simulation ***************************
   int n_e = 0;
   for (int n=1; n<max[i]; n++) {//Generate max[i] runs
          int r_e = generator.nextInt(100);
          if (r_e < 10) {                 // Machine 1
              int r_d = generator.nextInt(100);
              if (r_d < 5) {n_e++;}
          }
          else if (r_e < 45) {            // Machine 2
                  int r_d = generator.nextInt(100);
                  if (r_d < 3) {n_e++;}
          }
          else {                          // Machine 3
               int r_d = generator.nextInt(100);
               if (r_d < 1) {n_e++;}
         }
   }
   double prob = (double)n_e/max[i]; //Approximate probability
// ***  End Simulation ***************************
```

n	50	100	1,000	10,000	50,000	100,000	500,000	1,000,000
Prob	0.04	0.02	0.02	0.0227	0.02018	0.02082	0.021192	0.021018

Table 1.3: Probability that the selected item is defective. Exact answer = 0.0210

Exercise 1.8.2 Consider the problem of rolling three dice until six spots appear on all three of them. Simulate this scenario to estimate the probability that this will take at least 30 throws.

Answer 1.8.2 We first estimate the probability of getting three sixes in 29 or less throws. The probability of needing 30 or more throws is one minus this quantity.

```
// ***  Begin Simulation ***************************
         int n_e = 0;
         for (int n=1; n<max[i]; n++) {     //Generate max[i] runs
             int runs = 0; int tot = 0;
             while (runs < 29 && tot != 18) {
                runs++;
                int r1= generator.nextInt(6)+1;     // First die
                int r2= generator.nextInt(6)+1;     // Second die
                int r3= generator.nextInt(6)+1;     // Third die
                tot = r1+r2+r3;
                if (tot == 18) { n_e++; }
             }
         }
         double prob = (double)n_e/max[i]; //Approximate probability
// ***  End Simulation ***************************
```

One run of this simulation produced the following table of results.

n	50	100	1,000	10,000	50,000	100,000	500,000	1,000,000
Prob	0.12	0.08	0.126	0.1214	0.12586	0.12789	0.126054	0.125637

Table 1.4: Probability of getting three sixes in 29 or less throws.

Exercise 1.8.3 Consider the gambler's ruin problem in which a gambler begins with \$50, wins \$10 on each play with probability $p = 0.45$, or loses \$10 with probability $q = 0.55$. The gambler will quit once he doubles his money or has nothing left of his original \$50. Write a simulation program to compute the expected number of times the gambler has

(a) \$90 before ending up broke;
(b) \$90 before doubling his money;
(c) \$10 before doubling his money;
(d) \$10 before going broke.

Answer 1.8.3 The Java code and the results obtained with a single run of this code are as follows.

```
import java.util.Random;
class gamb {
   public static void main (String args[]) {
      Random generator = new Random();
      int max = 1000000;   int max1=0;   int max2 = 0;
      int avecount1b10 = 0; int avecount1b0 = 0;
      int avecount9b10 = 0; int avecount9b0 = 0;

      for (int n = 1; n<=max; n++) {            // Perform "max" trails

         int i = 5; int count1 = 0; int count9 = 0;
         do {                                   // Begin gambling
            int next = generator.nextInt(100);
            if (next < 55) {i--;}
            if (next >= 55) {i++;}
            if (i == 1 ) {count1++;}            // Down to last $10
            if (i == 9 ) { count9++;}           // Up to $90
         }
         while (i > 0 && i < 10);

         if (i == 0) {                          // Gambler ends up broke
            max1++;                   // Number of times he ends up broke
            avecount1b0 =avecount1b0+count1;//Update counts for averages
            avecount9b0 =avecount9b0+count9;
         }
         if (i == 10) {                       // Gambler doubles his money
            max2++;                      // Number of times he doubles up
            avecount1b10 =avecount1b10+count1; //Update counts for averages
            avecount9b10 =avecount9b10+count9;
         }
      }

      double average9b0  = (double) avecount9b0/max1;//Estimate mean values
      double average9b10 = (double) avecount9b10/max2;
      double average1b0  = (double) avecount1b0/max1;
      double average1b10 = (double) avecount1b10/max2;

System.out.println("In " + max + " tries each beginning with $50");
System.out.println("- number of times gambler ends up broke: " + max1);
System.out.println("- mean number of times he has $90 before going broke: "
+ average9b0);
System.out.println("- mean number of times he has $10 before going broke: "
+ average1b0);

System.out.println(" ");
System.out.println("- number of times he doubles his money: " + max2);
System.out.println("- mean number of times he has $90 before doubling up: "
```

```
 + average9b10);
System.out.println("- mean number of times he has $10 before doubling up: "
 + average1b10);
   }
}
```

The following output was obtained on one run of this program:

```
In 1000000 tries each beginning with $50
- number of times gambler ends up broke: 731888
- mean number of times he has $90 before going broke:
   0.17097561375511008
- mean number of times he has $10 before going broke:
   1.7547356972651553

- number of times he doubles his money: 268112
- mean number of times he has $90 before doubling up:
   1.7577803306081041
- mean number of times he has $10 before doubling up:
   0.16980590201109985
```

Exercise 1.8.4 By enclosing a circle of unit radius and centered at the origin inside a square box whose sides are of length 2, compute an approximation to the number π using a sequence of uniformly distributed random numbers.

Answer 1.8.4 Since the area of the square box is 4 and that of the circle is π, the percentage of the area of the box occupied by the circle is $\pi/4$. Thus we can generate random points (x_i, y_i) that are uniformly distributed over the square box and use the percentage that falls within the circle as an approximation to $\pi/4$. If r_1 and r_2 are random numbers that are uniformly distributed on $[0, 1]$, then $x_i = 2r_1 - 1$ and $y_i = 2r_2 - 1$ are uniformly distributed on $[-1, 1]$ and the point (x_i, y_i) is uniformly distributed over our box. The point (x_i, y_i) lies in the area occupied by the circle if $x_i^2 + y_i^2 \leq 1$. Our simulation must therefore generate a sequence of uniformly distributed random numbers between 0 and 1 and find the percentage of consecutive pairs that satisfy

$$(2r_1 - 1)^2 + (2r_2 - 1)^2 \leq 1$$

This percentage is taken as an approximation to $\pi/4$ and hence an approximation of π can be found. The following Java program does just this.

```
// ***  Begin Simulation ***************************
        int n_e = 0;
        for (int n=1; n<max[i]; n++) {          // Generate max[i] runs.

             double r1 = generator.nextDouble();
             double r2 = generator.nextDouble();
```

```
            if ( (2*r1-1)*(2*r1-1) + (2*r2-1)*(2*r2-1) <= 1) {n_e++;}
        }

        double prob = (double)n_e/max[i];  // Approximate probability
        double pi_app = prob*4;
        System.out.println( "Approximation to pi: " + pi_app);
// ***  End Simulation ***************************
```

A single run of this program produced the following results.

n	50	100	1,000	10,000	50,000	100,000	500,000	1,000,000
Prob	2.88	3.08	3.128	3.1364	3.14912	3.13544	3.138896	3.141064

Table 1.5: Estimations of π.

The exact answer is of course, $\pi = 3.14159$.

Chapter 2

Combinatorics — The Art of Counting

Exercise 2.1.1 For each of the following words, how many different arrangements of the letters can be found: *RECIPE, BOOK, COOKBOOK*?

Answer 2.1.1 The number of different arrangement of

(a) *RECIPE* is $6!/2! = 360$,
(b) *BOOK* is $4!/2! = 12$,
(c) *COOKBOOK* is $8!/(4!\,2!) = 840$.

Exercise 2.1.2 Find a word for which the number of different arrangements of its letters is equal to 30.

Answer 2.1.2 Since $30 = 5!/(2!\,2!)$, all 5-letter words containing two letters, each repeated twice, and one unique letter have 30 different arrangements. Some examples are:

DEEDS, MADAM, CIVIC, QUEUE, LEVEL, and so on.

Exercise 2.2.1 A pizza parlor offers ten different toppings on its individually sized pizzas. In addition, it offers thin crust and thick crust. William brings his friends Jimmy and Jon with him and each purchases a single one-topping pizza. How many different possible commands can be sent to the kitchen?

Answer 2.2.1 Each of the three friends can choose one of 20 different possibilities, one of ten toppings on either thin or thick crust. The number of different possibilities is 20^3.

Exercise 2.2.2 Billy and Kathie bring two of their children to a restaurant in France. The menu is composed of an entree, a main dish, and either dessert or cheese. The entree is chosen from among four different possibilities, one of which is snails. Five different possibilities, including two seafood dishes, are offered as the main dish. To the dismay of the children, who always choose dessert over cheese, only two possibilities are offered for dessert. The cheese plate is a restaurant-specific selection and no other choice of cheese is possible. Although Billy will eat anything, neither Kathie nor the children like snails and one of the children refuses to eat seafood. How many different possible commands can be sent to the kitchen?

Answer 2.2.2 Since Billy will eat anything, he has the choice of $4 \times 5 \times 3 = 60$ possibilities. Kathie, however, has only $3 \times 5 \times 3 = 45$ possibilities (since she will not eat snails). The child who will eat fish has $3 \times 5 \times 2 = 30$ possibilities and the other child has only $3 \times 3 \times 2 = 18$ possibilities. The total number of possible commands that may be sent to the kitchen is then $60 \times 45 \times 30 \times 18 = 1,458,000$.

Exercise 2.2.3 Consider a probability experiment in which four fair dice are thrown simultaneously. How many different outcomes (permutations) can be found? What is the probability of getting four 6's?

Answer 2.2.3 Let d_i, $i = 1, 2, 3, 4$, be the number of spots obtained by the ith die. In this sample space, each elementary event may be represented by a 4-tuple, (d_1, d_2, d_3, d_4), where, for each die d_i, we have $1 \le d_i \le 6$. It follows that there are $6^4 = 1,296$ equiprobable elementary events. Of all these elementary events, only one corresponds to the required answer, $(6, 6, 6, 6)$. Consequently, the probability of getting four sixes when four fair dice are thrown is $1/1296$.

Exercise 2.3.1 William Stewart's four children come down for breakfast at random times in the morning. What is the probability that they appear in order from the oldest to the youngest.
Note: William Stewart does not have twins (nor triplets, nor quadruplets!).

Answer 2.3.1 There are 4! different ways in which the children can appear for breakfast, and from oldest to youngest is only one of these ways. The probability is therefore given by $1/4! = 1/24$.

Exercise 2.3.2 The Computer Science Department at North Carolina State University wishes to market a new software product. The name of this package must consist of three letters only.

(a) How many possibilities does the department have to choose from?
(b) How many possibilities are there if exactly one vowel is included?
(c) How many possibilities are there if all three characters must be different and exactly one must be a vowel?

Answer 2.3.2

(a) $26^3 = 17,576$, since any of 26 letters may be chosen as the first, second, or third.

(b) Any of the five vowels may be placed in any of the three available slots. This gives 5×3 possibilities for the vowel. This leaves two slots for the consonants. The first can have any of 21 possibilities and idem for the second. The total number of choices is therefore $5 \times 3 \times 21 \times 21 = 6,615$.

(c) This only alters the choice of the second constant, whose choice must now be one of 20. The answer is $5 \times 3 \times 21 \times 20 = 6,300$.

Exercise 2.4.1 A box contains two white balls and four black ones.

(a) Two balls are chosen at random.
What is the probability that they are of the same color?
(b) Three balls are chosen at random.
What is the probability that all three are black?

Answer 2.4.1

(a) The sample space, which consists of combinations of two balls from among six, has $C(6,2) = 15$ elements. The event, $A =$ "two balls of the same color" can be separated into two disjoint events, namely,

A_1: Two white balls are chosen,
A_2: Two black balls are chosen.

There is $C(2,2) = 1$ way of choosing the two white balls and $C(4,2) = 6$ ways of choosing two black balls. Hence,

$$\text{Prob}\{A\} = \text{Prob}\{A_1\} + \text{Prob}\{A_2\} = \frac{1}{15} + \frac{6}{15} = \frac{7}{15}.$$

(b) This time the sample space contains $C(6,3) = 20$ elements. The number of ways in which three black balls can be chosen from among the four black balls is $C(4,3) = 4$. Hence, the answer is given as $4/20 = 1/5$.

Exercise 2.4.2 Yves Bouchet, guide de haute montagne, leads three clients on "la traversé de la Meije." All four spend the night in the refuge called "La Promotoire" and must set out on the ascension at 3:00 a.m., i.e., in the dark. Yves is the first to arise and, since it is pitch dark, randomly picks out two boots from the four pairs left by him and his clients.

(a) What is the probability that Yves actually gets his own boots?
(b) If Yves picks two boots, what is the probability that he chooses two left boots?
(c) What is the probability of choosing one left and one right boot?
(d) If, instead of picking just two boots, Yves picks three, what is the probability that he finds his own boots among these three?

Answer 2.4.2

(a) Yves picks two at random from among eight boots. The total number of ways to do this is $C(8,2) = 28$ and each is equally probable. Since only one of these combinations is the pair that belong to Yves, he has only one chance in 28 of picking the right pair. The requested probability is therefore $1/28$.

(b) The number of possible combinations is still $C(8,2) = 28$. The number of ways in which Yves can pick out two left boots is the number of combinations obtained in choosing 2 from 4, i.e., $C(4,2) = 6$. Thus the probability of choosing two left boots is $6/28$.

(c) In this case, Yves must choose one left boot from among four, ($C(4,1)$= 4) and one right boot from among four, ($C(4,1)$= 4). The probability of choosing one left and one right is therefore

$$\frac{4 \times 4}{28} = \frac{4}{7}.$$

(d) The number of ways of picking three boots from among eight is given as $C(8,3) = 56$. The number of possibilities that include both of Yves boots is 6; his own pair and one of any of the remaining six. The probability of finding his own among three picked at random is therefore $6/56 = 3/28$, i.e., three times more probable than that in choosing only two.

Exercise 2.4.3 A water polo team consists of seven players, one of whom is the goalkeeper, and six reserves who may be used as substitutes. The Irish water polo selection committee has chosen 13 players from which to form a team. These 13 include two goalkeepers, three strikers, four "heavies," i.e., defensive players, and four generalists, i.e., players who function anywhere on the team.
(a) How many teams can the coach form, given that he wishes to field a team containing a goalkeeper, one striker, two heavies, and three generalists?
(b) Toward the end of the last period, Ireland is leading its main rival, England, and, with defense in mind, the coach would like to field a team with all four heavies (keeping, of course, one goalkeeper, one striker, and one generalist). From how many teams can he now choose?

Answer 2.4.3

(a) One goalkeeper is chosen from two, one striker from three, two heavies from four and the remaining three players are chosen from four. This gives $C(2,1) \times C(3,1) \times C(4,2) \times C(4,3) = 144$ different teams that can be formed.

(b) Again, one goalkeeper is chosen from two, one striker from three, four heavies from four, and the seventh and last player is chosen from the four generalists. This gives $C(2,1) \times C(3,1) \times C(4,4) \times C(4,1) = 24$ possibilities.

Exercise 2.4.4 While in France, William goes to buy baguettes each morning and each morning he takes the exact change, 2.30 euros, from his mother's purse. One morning he finds a 2 euro coin, two 1 euro coins, five 0.20 euro coins, and four 0.10 euro coins in her purse. William has been collecting coins and observes that no coin of the same value comes from the same European country (i.e., all the coins are distinct). How many different possibilities has William from which to choose the exact amount?

Answer 2.4.4 If William chooses the 2-euro coin, then the remaining 0.30 can be chosen from three of the 0.10 coins, $C(4,3)$, or from one of the five 0.20 coins and one of the 0.10 coins $C(5,1) \times C(4,1)$. This gives a total of $C(4,3) + C(5,1) \times C(4,1) = 24$.

If William does not choose the 2-euro coin, then the two euro must be chosen as one of the following:

- the two 1-euro coins, in which case the remainder can be chosen in exactly any of the 24 different ways obtained when the 2-euro coin was picked.
- one of the two 1-euro coins and all five 0.20 coins. The remaining 0.30 is then selected from the four 0.10 coins in $C(4,3)$ ways. This gives a total of $2 \times C(4,3) = 8$ possibilities.

These are the only possibilities. The total number of combinations is therefore $24 + 24 + 8 = 56$.

Exercise 2.4.5 A box contains two white balls, two red balls, and a black ball. Balls are chosen without replacement from the box.

(a) What is the probability of choosing a red ball before the black ball?
(b) What is the probability of choosing the two white balls before choosing any other ball?

Answer 2.4.5

(a) We can neglect the fact that the box contains white balls; only the two red balls and the black ball enter into the analysis. The probability of choosing a red ball first is therefore $2/3$.

(b) There are $C(5,2) = 10$ ways in which two balls can be choosen, but only one way in which the two white balls can be chosen. The probability of choosing the two white balls is therefore $1/10$.

Alternatively, let A be the event "the first is white" and B be the event "the second is white." The required probability is then

$$\text{Prob}\{A \cap B\} = \text{Prob}\{A\}\text{Prob}\{B|A\} = \frac{2}{5} \times \frac{1}{4} = \frac{1}{10}.$$

Exercise 2.4.6 Three boxes, one large, one medium, and one small, contain white and red balls. The large box, which is chosen half the time, contains fifteen white balls and eight red ones; the medium box, which is chosen three times out of ten, contains nine white balls and three red ones. The small box contains four white balls and five red ones.

(a) On choosing two balls at random from the large box, what is the
   (i) probability of getting two white balls?
   (ii) getting one white and one red ball?

(b) After selecting a box according to the prescribed probabilities, what is the probability of getting one white and one red ball from this box?

(c) Given that a box is chosen according to the prescribed probabilities, and that a white ball is randomly picked from that box, what is the probability that the ball was actually chosen from the large box?

Answer 2.4.6 (a) The large box contains 23 balls. The number of ways in which two balls can be chosen is $C(23,2)$. The number of ways in which two white balls can be chosen from the 15 in this box is $C(15,2)$. Thus the probability of getting two white balls from the large box is

$$C(15,2)/C(23,2) = \frac{15 \times 7}{23 \times 11} = .415.$$

The probability of getting one of each color is

$$\frac{C(15,1) \times C(8,1)}{C(23,2)} = \frac{120}{253}.$$

(b) From the previous part, we know the probability of this event if the large box is chosen. If the medium box is chosen, the probability of getting two different colors is

$$\frac{C(9,1) \times C(3,1)}{C(12,2)} = \frac{9}{22}.$$

For the small box, this probability is

$$\frac{C(4,1) \times C(5,1)}{C(9,2)} = \frac{5}{9}.$$

The answer is therefore given by

$$\frac{1}{2} \times \frac{120}{253} + \frac{3}{10} \times \frac{9}{22} + \frac{2}{10} \times \frac{5}{9}.$$

(c) We use Bayes' law to obtain

$$\text{Prob}\{B|W\} =$$

$$\frac{\text{Prob}\{B\}\text{Prob}\{W|B\}}{\text{Prob}\{B\}\text{Prob}\{W|B\} + \text{Prob}\{M\}\text{Prob}\{W|M\} + \text{Prob}\{S\}\text{Prob}\{W|S\}} =$$

$$\frac{1/2 \times 15/23}{1/2 \times 15/23 + 3/10 \times 9/12 + 2/10 \times 4/9}$$

where B = Big; M = Medium; S = Small; W = White and R = Red.

Exercise 2.4.7 Consider a system that consists of p processes that have access to r identical units of a resource. Suppose each process alternates between using a single unit of the resource and not using any resource.

(a) Describe a sample space that illustrates the situation of the p processes.

(b) What is the size of this sample space when (a) $p = 6,\ r = 8$ and (b) $p = 6,\ r = 4$?

(c) In this first case ($p = 6,\ r = 8$), define the following events.

- $\mathcal{A}$: Either two or three processes are using the resource.
- $\mathcal{B}$: At least three but not more than five processes are using the resource.
- $\mathcal{C}$: Either all the processes are using the resource or none are.

Assuming that each process is equally likely as not to be using a unit of resource, compute the probabilities of each of these three events.

(d) Under the same assumption, what do the following events represent, and what are their probabilities?

- $\mathcal{A} \cap \mathcal{B}$.
- $\mathcal{A} \cup \mathcal{B}$.
- $\mathcal{A} \cup \mathcal{B} \cup \mathcal{C}$.

Answer 2.4.7

(a) An element of the sample space may be written as a vector of length p, the ith element of which denotes the status of the ith process. If the ith element, p_i say, is equal to 0, then process i is not using a unit of resource; if it is equal to 1, then process i is using a unit of resource. The sample space is then the set of all vectors of length p, whose components satisfy the conditions $p_i \in \{0, 1\}$ and $\sum_{i=1}^{p} p_i \leq r$.

(b) For example, with $p = 6$ and $r = 8$, some elements of the sample space are $(0,0,0,0,0,0)$, which represents the elementary event corresponding to the situation in which no process is using a resource; $(0,1,0,1,1,0)$, the elementary event in which processes 2, 4 and 5 are using the resource, and processes 1, 3 and 6 are not; and finally $(1,1,1,1,1,1)$, the elementary event representing the case when all processes are using a unit of resource. In this case ($p = 6$ and $r = 8$) all processes can have a unit of resource simultaneously. The total number of elementary events is then given by $2^6 = 64$.

Another way to observe this is to realize that the number of ways in which exactly k processes, $k \leq p$, can access the resource is just the number of ways of choosing k from p, i.e., $C(p,k) = p!/(p-k)!k!$. We now allow k to vary from 0 to p and the total number of elements in the sample space is then given by

$$\sum_{k=0}^{p} C(p,k).$$

With $p = 6$, we have

$$\begin{array}{llll} \mathrm{C(6,0)} = 1; & \mathrm{C(6,1)} = 6; & \mathrm{C(6,2)} = 15; & \mathrm{C(6,3)} = 20; \\ \mathrm{C(6,4)} = 15; & \mathrm{C(6,5)} = 6; & \mathrm{C(6,6)} = 1. & \end{array}$$

Hence the total number is given by $1 + 6 + 15 + 20 + 15 + 6 + 1 = 64$.

When the number of units of resource is strictly less than the number of processes, $p > r$, not all processes can simultaneously access units of resource so some of the elementary events of the previous case will not occur. When $r = 4$, elementary events corresponding to $C(6,5)$ and $C(6,6)$ need to be removed. The number of elementary events is then given by $1 + 6 + 15 + 20 + 15 = 57$.

(c)
- Prob$\{A\} = \frac{C(6,2)+C(6,3)}{64} = 35/64$.
- Prob$\{B\} = \frac{C(6,3)+C(6,4)+C(6,5)}{64} = 41/64$.
- Prob$\{C\} = \frac{C(6,0)+C(6,6)}{64} = 2/64$.

(d) Given the events A, B and C defined in the question, it is easy to observe that

- Prob$\{A \cap B\}$ = Prob{Exactly 3 processes are using the resource} $= 20/64 = 0.625$.
- Prob$(A \bigcup B)$ = Prob{2, 3, 4 or 5 processes are using the resource} $= 15/64 + 20/64 + 15/64 + 6/64 = 7/8 = 0.875$.
- Prob$(A \bigcup B \bigcup C)$ = Prob{0, 2, 3, 4, 5 or 6 processes are using the resource} $= 1 - 6/64 = 0.90625$.

Exercise 2.5.1 Show that, as the population size n grows large, the difference between the number of combinations *without* replacement for a fixed number of items k becomes equal to the number of combinations *with* replacement.

Answer 2.5.1 The ratio of these two quantities is given by

$$\frac{(n+k-1)!}{k!\,(n-1)!} \bigg/ \frac{n!}{k!\,(n-k)!}$$

$$= \frac{(n+k-1)!}{n!}\frac{(n-k)!}{(n-1)!} = \frac{(n+k-1)(n+k-2)\cdots(n+1)}{(n-1)(n-2)\cdots(n-k+1)}.$$

In this quotient, observe that the number of terms on the top is equal to the number on the bottom: both have $k-1$ terms. Taking them one at a time, we see that the ratio is given by

$$\frac{n+k-1}{n-1} \times \frac{n+k-2}{n-2} \times \cdots \times \frac{n-k+1+k}{n-k+1}$$

$$= \left(1+\frac{k}{n-1}\right)\left(1+\frac{k}{n-2}\right)\left(1+\frac{k}{n-(k-1)}\right),$$

which tends to 1 as $n \to \infty$, for any fixed value of k.

Exercise 2.6.1 The probability that a message is successfully transmitted over a communication channel is known to be p. A message that is not received correctly is retransmitted until such time as it is received correctly. Assuming that successive transmissions are independent, what is the probability that no retransmissions are needed? What is the probability that exactly two retransmissions are needed?

Answer 2.6.1 Let $q = 1 - p$. The probability of having no retransmissions is equal to the probability that the message was successfully transmitted the first time, i.e.,

$$\text{Prob}\{\text{No retransmissions}\} = p.$$

The probability of having two transmissions is equal to the probability that the message was unsuccessfully sent the first two times (q^2), but was successful on the third try. Thus

$$\text{Prob}\{\text{Two retransmissions exactly}\} = (1-p)^2 p.$$

Exercise 2.6.2 The outcome of a Bernoulli trial may be any one of m possibilities which we denote $h_1,\ h_2, \ldots, h_m$. Given n positions, show that the number of ways of placing h_1 in n_1 positions, h_2 in n_2 positions and so on, with $\sum_{j=1}^{m} n_j = n$, is given by the binomial coefficient,

$$\binom{n}{n_1, n_2, \cdots, n_m} = \frac{n!}{n_1! n_2! \cdots n_m!}.$$

Answer 2.6.2 The number of ways of placing h_1 in n_1 out of a total of n positions is given by the binomial coefficient and is equal to

$$\frac{n!}{n_1!\,(n-n_1)!}.$$

This leaves $n-n_1$ positions into which h_2 must be placed into n_2 of them. Again, this is given by the binomial coefficient and is equal to

$$\frac{(n-n_1)!}{n_2!\,(n-n_1-n_2)!}.$$

Hence, the number of ways of placing h_1 into n_1 positions and h_2 into n_2 positions is

$$\frac{n!}{n_1!\,(n-n_1)!} \times \frac{(n-n_1)!}{n_2!\,(n-n_1-n_2)!} = \frac{n!}{n_1!\,n_2!\,(n-n_1-n_2)!}.$$

This now leaves $n-n_1-n_2$ positions and h_3 is to be placed in n_3 of them. The number of ways in which this may be accomplished is given by

$$\frac{(n-n_1-n_2)!}{n_3!\,(n-n_1-n_2-n_3)!}$$

and hence the number of ways of satisfying the requirements for h_1, h_2 and h_3 is

$$\frac{n!}{n_1!\,(n-n_1)!} \times \frac{(n-n_1)!}{n_2!\,(n-n_1-n_2)!} \times \frac{(n-n_1-n_2)!}{n_3!\,(n-n_1-n_2-n_3)!}$$

$$= \frac{n!}{n_1!\,n_2!\,n_3!\,(n-n_1-n_2-n_3)!}.$$

The process terminates when the term in parenthesis is equal to zero, i.e., the point at which all m items have been placed. This gives the multinomial coefficient,

$$\frac{n!}{n_1!\,n_2!\,n_3!\,\cdots\,n_m!} = \binom{n}{n_1, n_2, \cdots, n_m}.$$

Chapter 3

Random Variables and Distribution Functions

Exercise 3.1.1 A four-sided die (a tetrahedron having the numbers 1, 2, 3, and 4 printed on it, one per side) is thrown twice. Let M be the random variable that denotes the maximum number obtained on the two throws. What is the domain and range of M? Give the partition of the state space induced by this random variable. What is the set A_3?

Answer 3.1.1 The domain of M is the set of all (i, j) for $i = 1, 2, 3, 4$ and $j = 1, 2, 3, 4$. It is the set of consisting of the elementary events of the sample space. The range of M is the set $\{1, 2, 3, 4\}$. The partition induced by M is as follows:

$$\begin{aligned} (1,1) &\Rightarrow 1 \\ (1,2),\ (2,2),\ (2,1) &\Rightarrow 2 \\ (1,3),\ (2,3),\ (3,3),\ (3,2),\ (3,1) &\Rightarrow 3 \\ (1,4),\ (2,4),\ (3,4),\ (4,4),\ (4,3),\ (4,2),\ (4,1) &\Rightarrow 4 \end{aligned}$$

The set A_3 is the set of outcomes for which the maximum value is 3, i.e.,

$$A_3 = \{(1,3),\ (2,3),\ (3,3),\ (3,2),\ (3,1)\}.$$

Exercise 3.1.2 A survey reveals that 20% of students at Tobacco Road High School smoke while 15% drink beer, both considered to be bad habits. Let X be the random variable that denotes the number of these bad habits indulged by a randomly chosen student. What is the domain and range of X? Describe the partition of the state space induced by this random variable.

Answer 3.1.2 The domain of X is the set of all students in the school. Its range is the set of three integers $\{0, 1, 2\}$ since any given student may have none, one or both vices. The random variable X partitions the students into those who have neither bad habits; one or the other of the two but not both, and those students who have both.

Exercise 3.2.1 Let X be a random variable whose probability mass function is given as

$$p_X(x) = \begin{cases} \alpha/x, & \text{x} = 1,2,3,4, \\ 0 & \text{otherwise.} \end{cases}$$

(a) What is the value of α?

(b) Compute Prob$\{X \text{ is odd}\}$.

(c) Compute Prob$\{X > 2\}$.

Answer 3.2.1

(a) Since $\sum_{x=1,2,3,4} p_X(x) = 1$, we must have $\alpha(1+1/2+1/3+1/4) = \alpha(25/12) = 1$ which implies that $\alpha = 12/25$.

(b) Prob$\{X \text{ is odd}\} = p_X(1) + p_X(3) = 12/25 \times (1+1/3) = 16/25$.

(c) Prob$\{X > 2\} = p_X(3) + p_X(4) = 12/25 \times (1/3+1/4) = 7/25$.

Exercise 3.2.2 Balls are drawn from an urn containing two white balls and five black balls until a white ball appears. Let X be the random variable that denotes the number of black balls drawn *before* a white ball appears. What is the domain and range of X and what is the partition induced on the sample space by X? Give a table of the probability mass function of X.

Answer 3.2.2 The domain of X (its sample space) contains six elements, namely

$$\{w, bw, bbw, bbbw, bbbbw, bbbbbw\}$$

where the letter b at any position $i = 1,2,3,4,5$ indicates that a black ball is chosen at step i and w that a white ball is chosen. The range of X is the set $\{0,1,2,3,4,5\}$

Observe that the probability that a white ball is drawn from a total of w white balls and b black balls is given by $w/(b+w)$ while the probability of choosing a black ball in these same circumstances is $b/(b+w)$. The probability that no black ball is drawn before a white ball is given by 2/7; the probability that one black ball is drawn before a white ball is $5/7 \times 2/6$; the probability that two black balls are drawn before a white ball is $5/7 \times 4/6 \times 2/5$, and so on. The probability mass function of X is given by

0	1	2	3	4	5
$\frac{2}{7}$	$\frac{5}{7} \times \frac{2}{6}$	$\frac{5}{7} \times \frac{4}{6} \times \frac{2}{5}$	$\frac{5}{7} \times \frac{4}{6} \times \frac{3}{5} \times \frac{2}{4}$	$\frac{5}{7} \times \frac{4}{6} \times \frac{3}{5} \times \frac{2}{4} \times \frac{2}{3}$	$\frac{5}{7} \times \frac{4}{6} \times \frac{3}{5} \times \frac{2}{4} \times \frac{1}{3} \times \frac{2}{2}$

or

0	1	2	3	4	5
$\frac{6}{21}$	$\frac{5}{21}$	$\frac{4}{21}$	$\frac{3}{21}$	$\frac{2}{21}$	$\frac{1}{21}$

Observe the pattern in these probabilities as well as the fact that their sum is equal to one.

Exercise 3.2.3 An urn contains seven white balls and five black ones. Suppose n balls are chosen at random. Let the random variable X denote the number of white balls in the sample. What is the probability mass function of X if the n balls are chosen (a) without replacement? (b) with replacement?

Answer 3.2.3

(a) Without replacement, the total number of ways in which n balls can be chosen from twelve is $C(12, n)$. If among n balls, k of them are white, then the number of ways in which k white balls can be chosen from among seven is $C(7, k)$, assuming that $k \leq 7$. The number of ways in which the remaining $n - k$ black balls can be chosen from five is $C(5, n - k)$. Thus

$$\text{Prob}\{X = k\} = \frac{C(7,k) \times C(5, n-k)}{C(12, n)} \quad \text{for } 0 \leq k \leq \min(n, 7).$$

With $n = 8$, we obtain the following values:

$$\begin{aligned}
&\text{Prob}\{X = 3\} = 35/495 \approx 0.07, \\
&\text{Prob}\{X = 4\} = 175/495 \approx 0.35, \\
&\text{Prob}\{X = 5\} = 210/495 \approx 0.42, \\
&\text{Prob}\{X = 6\} = 70/495 \approx 0.14, \text{ and,} \\
&\text{Prob}\{X = 7\} = 5/495 \approx 0.01.
\end{aligned}$$

(b) With replacement, it can be viewed as a Bernoulli trial, then the probability is given by

$$C(n, k)\left(\frac{7}{12}\right)^k \left(\frac{5}{12}\right)^{n-k} \quad \text{for } 0 \leq k \leq n.$$

With $n = 8$, we obtain the following values:

$$\begin{aligned}
&\text{Prob}\{X = 0\} = C(8, 0)(\tfrac{7}{12})^0(\tfrac{5}{12})^8 \approx 0.0009, \\
&\text{Prob}\{X = 1\} \approx 0.01, \\
&\text{Prob}\{X = 2\} \approx 0.05, \\
&\text{Prob}\{X = 3\} \approx 0.14,
\end{aligned}$$

$$\begin{aligned}
&\text{Prob}\{X=4\}\approx 0.24,\\
&\text{Prob}\{X=5\}\approx 0.27,\\
&\text{Prob}\{X=6\}\approx 0.19,\\
&\text{Prob}\{X=7\}\approx 0.07, \text{ and,}\\
&\text{Prob}\{X=8\}\approx 0.01.
\end{aligned}$$

Exercise 3.3.1 The cumulative distribution function of a discrete random variable X is given as

$$F_X(x)=\begin{cases} 0, & x<0,\\ 1/4, & 0\le x<1,\\ 1/2, & 1\le x<2,\\ 1, & x\ge 2.\end{cases}$$

Find the probability mass function of X.

Answer 3.3.1 Observe that this random variable has discontinuities at the points $x = 0, 1$ and 2. These are the values that X can assume and are the only points at which $p_X(x)$ can have a non-zero value. By subtracting the value of the function just prior to one of these points from the value of the function at the point itself, we obtain the following probability mass function of X.

$$p_X(x)=\begin{cases} 1/4, & x=0,\\ 1/4, & x=1,\\ 1/2, & x=2,\\ 0 & \text{otherwise.}\end{cases}$$

Exercise 3.3.2 The cumulative distribution function of a discrete random variable N is given as

$$F_N(x)=1-\frac{1}{2^n} \quad \text{if } x\in[n,n+1) \quad \text{for } n=1,2,\ldots$$

and has the value 0 if $x<1$.

(a) What is the probability mass function of N?

(b) Compute $\text{Prob}\{4\le N\le 10\}$.

Answer 3.3.2 Observe that this random variable has discontinuities at all the points $x = 1, 2, \ldots$ and that

$$F_N(1)=1/2,\quad F_N(2)=3/4,\quad F_N(3)=7/8,\ldots$$

and approaches 1 as $n\to\infty$.

(a) The probability mass function of N is

$$p_N(n) = \begin{cases} 1/2^n, & n = 1, 2, \ldots, \\ 0 & \text{otherwise.} \end{cases}$$

(b)

$$\text{Prob}\{4 \leq N \leq 10\} = F_N(10) - F_N(3) = \left(0.5^3 - 0.5^{10}\right) = \frac{1}{8} - \frac{1}{1024}$$
$$= 127/1024.$$

Exercise 3.3.3 A bag contains two fair coins and a third coin which is biased: the probability of tossing a head on this third coin is 3/4. A coin is pulled at random and tossed three times. Let X be the random variable that counts the number of heads obtained in these three tosses. Previously, we found the probability mass function of X. Now find its the cumulative density function.

Answer 3.3.3

Previously, we found the table of the probability mass function of X to be

x_i	0	1	2	3
$p_X(x_i)$	$2/3 \times 1/8$ + $1/3 \times 1/64$	$2/3 \times 3/8$ + $1/3 \times 9/64$	$2/3 \times 3/8$ + $1/3 \times 27/64$	$2/3 \times 1/8$ + $1/3 \times 27/64$
$p_X(x_i)$	17/192	57/192	75/192	43/192

Since the random variable has nonzero value only at the four points 0, 1, 2 and 3, the cumulative distribution function of X is

$$F_X(x) = \begin{cases} 0, & x < 0, \\ 17/192, & 0 \leq x < 1, \\ 74/192, & 1 \leq x < 2, \\ 149/192, & 2 \leq x < 3, \\ 1, & x \geq 3. \end{cases}$$

Exercise 3.3.4 Let X be a continuous random variable whose cumulative distribution function is

$$F_X(x) = \begin{cases} 0, & x < 0, \\ x/5, & 0 \leq x \leq 5, \\ 1, & x > 5. \end{cases}$$

Compute the following probabilities:

(a) $\text{Prob}\{X \leq 1\}$.
(b) $\text{Prob}\{X > 3\}$.
(c) $\text{Prob}\{2 < X \leq 4\}$.

Answer 3.3.4 We have

(a) $\text{Prob}\{X \leq 1\} = F_X(1) = 1/5$,
(b) $\text{Prob}\{X > 3\} = 1 - \text{Prob}\{X \leq 3\} = 1 - F_X(3) = 1 - 3/5 = 2/5$,
(c) $\text{Prob}\{2 < X \leq 4\} = F(4) - F(2) = 4/5 - 2/5 = 2/5$.

Exercise 3.4.1 The cumulative distribution function of a continuous random variable X is given by

$$F_X(x) = \begin{cases} 0, & x < 0, \\ x^2/4, & 0 \leq x \leq 2, \\ 1, & x > 2. \end{cases}$$

Find the probability density function of X.

Answer 3.4.1 By differentiation, we obtain

$$f_X(x) = \begin{cases} x/2, & 0 \leq x \leq 2, \\ 0 & \text{otherwise.} \end{cases}$$

Exercise 3.4.2 The cumulative distribution function of a continuous random variable E is given by

$$F_E(x) = \begin{cases} 0, & x < 0, \\ 1 - e^{-\mu x} - \mu x e^{-\mu x}, & 0 \leq x < \infty, \end{cases}$$

and $\mu > 0$. Find the probability density function of E.

Answer 3.4.2 By differentiation, we obtain

$$\frac{d}{dx}F_E(x) = \mu e^{-\mu x} - \left(\mu x(-\mu e^{-\mu x}) + \mu e^{-\mu x}\right) = \mu^2 x e^{-\mu x}.$$

Thus, the probability density function of E is

$$f_E(x) = \begin{cases} 0, & x < 0, \\ \mu^2 x e^{-\mu x}, & 0 \leq x < \infty. \end{cases}$$

Exercise 3.4.3 The probability density function for a continuous "Rayleigh" random variable X is given by

$$f_X(x) = \begin{cases} \alpha^2 x e^{-\alpha^2 x^2/2}, & x > 0, \\ 0 & \text{otherwise.} \end{cases}$$

Find the cumulative distribution of X.

Answer 3.4.3 We have

$$F_X(x) = \int_0^x \alpha^2 t e^{-\alpha^2 t^2/2} dt = -e^{-\alpha^2 t^2/2}\Big|_0^x = 1 - e^{-\alpha^2 x^2/2}, \quad \text{if } x \geq 0,$$

and is zero otherwise.

Exercise 3.4.4 Let $f_1(x)$, $f_2(x)$, $\ldots, f_n(x)$ be a set of n probability density functions and let $p_1, p_2, \ldots, p_n$ be a set of probabilities for which $\sum_{i=1}^n p_i = 1$. Prove that $\sum_{i=1}^n p_i f_i(x)$ is a probability density function.

Answer 3.4.4 Since each of the $f_i(x)$ is a probability density function, we have, for $i = 1, 2, \ldots, n$,

$$f_i(x) \geq 0 \quad \text{for all} \quad x \in \Re,$$

and

$$\int_{-\infty}^{\infty} f_i(x)dx = 1.$$

From the first condition and the fact that $p_i \geq 0$ for all i, it immediately follows that

$$p_1 f_1(x) + p_2 f_2(x) + \cdots + p_n f_n(x) \geq 0 \quad \text{for all} \quad x \in \Re.$$

Also,

$$\begin{aligned} & \int_{-\infty}^{\infty} \left[p_1 f_1(x) + p_2 f_2(x) + \cdots + p_n f_n(x)\right] dx \\ = & \int_{-\infty}^{\infty} p_1 f_1(x)dx + \int_{-\infty}^{\infty} p_2 f_2(x)dx + \cdots + \int_{-\infty}^{\infty} p_n f_n(x)dx \\ = & \; p_1 \int_{-\infty}^{\infty} f_1(x)dx + p_2 \int_{-\infty}^{\infty} f_2(x)dx + \cdots + p_n \int_{-\infty}^{\infty} f_n(x)dx = \sum_{i=1}^n p_i = 1, \end{aligned}$$

which completes the proof.

Exercise 3.5.1 The cumulative distribution function of a discrete random variable X is given as

$$F_X(x) = \begin{cases} 0, & x < 0, \\ 1/4, & 0 \leq x < 1, \\ 1/2, & 1 \leq x < 2, \\ 1, & x \geq 2. \end{cases}$$

Find the probability mass function and the cumulative distribution function of $Y = X^2$.

Answer 3.5.1 The probability mass function of X is

$$p_X(x) = \begin{cases} 1/4, & x = 0, \\ 1/4, & x = 1, \\ 1/2, & x = 2, \\ 0 & \text{otherwise.} \end{cases}$$

The random variable X assumes values 0, 1 and 2 which implies that the random variable Y assumes the values 0, 1, and 4. Since this problem falls into the case that when $x_1 \neq x_2$ we have $f_X(x_1) \neq f_X(x_2)$, we have immediately

$$p_Y(y) = \begin{cases} 1/4, & y = 0, \\ 1/4, & y = 1, \\ 1/2, & y = 4, \\ 0 & \text{otherwise.} \end{cases}$$

From the probability mass function, we may form the cumulative distribution function as

$$F_Y(y) = \begin{cases} 0, & y < 0, \\ 1/4, & 0 \leq y < 1, \\ 1/2, & 1 \leq y < 4, \\ 1, & y \geq 4. \end{cases}$$

Exercise 3.5.2 A discrete random variable X assumes each of the values of the set $\{-10, -9, \ldots, 9, 10\}$ with equal probability. In other words, X is a discrete integer-valued random variable that is uniformly distributed on the interval $[-10, 10]$. Compute the following probabilities:

$\text{Prob}\{4X \leq 2\}, \text{Prob}\{4X + 4 \leq 2\}, \text{Prob}\{X^2 - X \leq 3\}, \text{Prob}\{|X - 2| \leq 2\}$

Answer 3.5.2

$$\begin{aligned} \text{Prob}\{4X \leq 2\} &= \text{Prob}\{X \leq 1/2\} = 11/21, \\ \text{Prob}\{4X + 4 \leq 2\} &= \text{Prob}\{X \leq -1/2\} = 10/21, \\ \text{Prob}\{X^2 - X \leq 3\} &= \text{Prob}\{X = -1\} + \text{Prob}\{X = 0\} + \text{Prob}\{X = 1\} \\ &\quad + \text{Prob}\{X = 2\} = 4/21, \\ \text{Prob}\{|X - 2| \leq 2\} &= \text{Prob}\{-2 \leq X - 2 \leq 2\} = \text{Prob}\{0 \leq X \leq 4\} \\ &= 5/21. \end{aligned}$$

Exercise 3.5.3 Homeowners in whose homes wildlife creatures take refuge call "Critter Control" to rid them of these "pests." Critter Control charges \$25 for each animal trapped and disposed of as well as a flat fee of \$45 per visit. Experience has shown that the distribution of trapped animals found per visit is as follows.

x_i	0	1	2	3	4
$p_X(x_i)$	0.5	0.25	0.15	0.05	0.05

Find the probability mass function of the amount of money Critter Control collects per visit.

Answer 3.5.3 Let Y be the random variable that describes the amount of money collected by Critter Control per visit. Then $Y = 25X + 45$ dollars. The probability mass function of Y is given as

y_i	45	70	95	120	145
$p_Y(y_i)$	0.5	0.25	0.15	0.05	0.05

Exercise 3.5.4 Let the probability distribution function of a continuous random variable X be given by

$$F_X(x) = \begin{cases} 1 - e^{-2x}, & 0 < x < \infty, \\ 0 & \text{otherwise.} \end{cases}$$

Find the cumulative distribution function of $Y = e^X$.

Answer 3.5.4 Using the relationship $e^{\ln y} = y$, we obtain

$$\begin{aligned} F_Y(y) = \text{Prob}\{Y \le y\} &= \text{Prob}\{e^X \le y\} \\ &= \text{Prob}\{X \le \ln y\} \\ &= F_X(\ln y) \\ &= 1 - e^{-2\ln y} = 1 - y^{-2}, \quad \text{for } 1 < y < \infty. \end{aligned}$$

Exercise 3.6.1 The probability mass function of a discrete integer-valued random variable is

$$p_X(x) = \begin{cases} 1/4, & x = -2, \\ 1/8, & x = -1, \\ 1/8, & x = 0, \\ 1/2, & x = 1, \\ 0 & \text{otherwise.} \end{cases}$$

Find $p_{X|\mathcal{B}}(x)$ where $\mathcal{B} = [X < 0]$.

Answer 3.6.1 We must first compute the probability of the event $\mathcal{B}$. From the probability mass function, we obtain

$$\text{Prob}\{\mathcal{B}\} = \text{Prob}\{X < 0\} = 3/8.$$

Then

$$p_{X|\mathcal{B}}(x) = \frac{p_X(x)}{\text{Prob}\{\mathcal{B}\}} = \frac{8}{3} p_X(x) \quad \text{if } x \in \mathcal{B}$$

and is zero otherwise. Thus

$$p_{X|\mathcal{B}}(x) = \begin{cases} 8/3 \times 1/4 = 2/3, & x = -2, \\ 8/3 \times 1/8 = 1/3, & x = -1, \\ 0 & \text{otherwise.} \end{cases}$$

Exercise 3.6.2 The probability mass function of a discrete random variable is given by

$$p_X(x) = \begin{cases} \alpha/x, & x = 1, 2, 3, 4, \\ 0 & \text{otherwise.} \end{cases}$$

Find $p_{X|\mathcal{B}}(x)$ where $\mathcal{B} = [X \text{ is odd}]$.

Answer 3.6.2 From a previous question, we have $\alpha = 12/25$ and $\text{Prob}\{X \text{ is odd}\} = 16/25$. Therefore

$$p_X(1) = 12/25 \quad \text{and} \quad p_X(3) = 4/25.$$

and

$$p_{X|\mathcal{B}}(x) = \frac{p_X(x)}{\text{Prob}\{\mathcal{B}\}} = \frac{25}{16} p_X(x) \quad \text{if } x \in \mathcal{B},$$

i.e.,

$$p_{X|\mathcal{B}}(x) = \begin{cases} 25/16 \times 12/25 = 3/4, & x = 1, \\ 25/16 \times 4/25 = 1/4, & x = 3, \\ 0 & \text{otherwise.} \end{cases}$$

Exercise 3.6.3 A biased coin shows heads with probability p and tails with probability $q = 1 - p$. The length of a *run of heads* is defined as the number of consecutive heads that appear; the length of a *run of tails* is similarly defined. Obviously runs of heads follow runs of tails which in turn follow runs of heads and so on forever. Let X_1 be the length of the first run and X_2 be the length of the second. What are the probability mass functions of X_1 and X_2.

Answer 3.6.3 If the first toss results in a head, then the probability that the very first run (a run of heads) will have length n is equal to $p^{n-1}q$, i.e., $\text{Prob}\{X_1 = n\} = p^{n-1}q$. On the other hand, if the first toss gives a tail, then $\text{Prob}\{X_1 = n\} = q^{n-1}p$. Observe that these are conditional probabilities and so we conclude

$$\begin{aligned} \text{Prob}\{X_1 = n\} &= \text{Prob}\{X_1 = n|\text{Head first}\}p + \text{Prob}\{X_1 = n|\text{Tail first}\}q \\ &= (p^{n-1}q)p + (q^{n-1}p)q = p^n q + q^n p, \quad \text{for } n = 1, 2, \ldots \end{aligned}$$

The results for X_2 are

$$\begin{aligned} \text{Prob}\{X_2 = n\} &= \text{Prob}\{X_2 = n|\text{Head first}\}p + \text{Prob}\{X_2 = n|\text{Tail first}\}q \\ &= (q^{n-1}p)p + (p^{n-1}q)q = q^{n-1}p^2 + p^{n-1}q^2, \text{ for } n = 1, 2, \ldots \end{aligned}$$

Chapter 4

Joint Random Variables and their Distributions

Exercise 4.1.1 Give an example (possibly from your own experience) of an instance in which two discrete random variables are jointly needed to represent a given scenario.

Answer 4.1.1 Possible Answer: Let X be the discrete random variable that describes the number of people severely injured in a recent earthquake and requiring hospitalization. Let Y be the random variable that described the number of available hospital beds in the town nearest to the disaster. Then $X - Y$ is the discrete random variable that describes the number of injured that must be transported to a hospital in a different town.

Exercise 4.1.2 Give an example (possibly from your own experience) of an instance in which two random variables, one discrete and one continuous, are jointly needed to represent a given scenario.

Answer 4.1.2 Possible Answer: Let X be a continuous random variable that denotes the acreage a Christmas Tree farmer has under cultivation, and let Y be a discrete random variable that counts the number of trees he sells per year. The farmer will likely make use of both X and Y in making decisions concerning which trees to cut in a particular year.

Exercise 4.2.1 Let X, Y, and Z be three discrete random variables for which

$$\begin{aligned}
\text{Prob}\{X=0,\ Y=0,\ Z=0\} &= 6/24,\\
\text{Prob}\{X=0,\ Y=1,\ Z=0\} &= 8/24,\\
\text{Prob}\{X=0,\ Y=1,\ Z=1\} &= 6/24,\\
\text{Prob}\{X=1,\ Y=0,\ Z=-1\} &= 1/24,\\
\text{Prob}\{X=1,\ Y=0,\ Z=1\} &= 1/24,\\
\text{Prob}\{X=1,\ Y=1,\ Z=0\} &= 2/24.
\end{aligned}$$

Let S be a new random variable for which $S = X + Y + Z$.

(a) Find the marginal probability mass function of X.

(b) Are X and Y independent?

(c) Are X and Z independent?

(d) Find the marginal probability mass function of S.

Answer 4.2.1

(a) The marginal probability mass functions of X, Y and Z respectively are

$X = 0$	$X = 1$
20/24	4/24

$Y = 0$	$Y = 1$
8/24	16/24

$Z = 0$	$Z = 1$	$Z = -1$
16/24	7/24	1/24

(b) X and Y are not independent since

$$6/24 = \text{Prob}\{X = 0,\ Y = 0\} \neq \text{Prob}\{X = 0\}\text{Prob}\{Y = 0\} = 160/24^2.$$

(c) X and Z are not independent since

$$6/24 = \text{Prob}\{X = 0,\ Z = 1\} \neq \text{Prob}\{X = 0\}\text{Prob}\{Z = 1\} = 140/24^2.$$

(d) The marginal probability mass functions of S is

$S = 0$	$S = 1$	$S = 2$
7/24	8/24	9/24

Exercise 4.2.2 Let X, Y, and Z be three discrete random variables for which

$$\begin{aligned}
\text{Prob}\{X = 1,\ Y = 1,\ Z = 0\} &= p, \\
\text{Prob}\{X = 1,\ Y = 0,\ Z = 1\} &= (1 - p)/2, \\
\text{Prob}\{X = 0,\ Y = 1,\ Z = 1\} &= (1 - p)/2,
\end{aligned}$$

where $0 < p < 1$. What is the joint probability mass function of X and Y?

Answer 4.2.2 The joint probability mass probability mass function is given by

	$X=0$	$X=1$
$Y=0$	0	$(1-p)/2$
$Y=1$	$(1-p)/2$	p

Exercise 4.2.3 Let X be a random variable that has the value $\{-1,\ 0,\ 1\}$ depending upon whether a child in school is performing below, at, or above grade level, respectively. Let Y be a random variable that is equal to zero if a child comes from an impoverished family and equal to one otherwise. In a particular class, it is observed that 20% of the children come from impoverished families and are performing below grade level, that 20% are from impoverished families and are performing at grade level, and that 6% are from impoverished families and are performing above grade level. Of the remaining children, half are performing at grade level and one-third are performing above grade level. Construct a table of the joint probability mass function of X and Y and compute the marginal probability density function of both random variables. Are X and Y independent?

Answer 4.2.3 Since 46% of the children come from impoverished families, 54% do not and of these half, i.e., 27% are performing at grade level. Similarly 18% of the children are not from impoverished families and are performing above grade level, the remaining 9% are not from impoverished families and are performing below grade level. The joint probability density function table, and the marginals are given below.

	$X=-1$	$X=0$	$X=1$	$p_Y(y)$
$Y=0$	0.20	0.20	0.06	0.46
$Y=1$	0.09	0.27	0.18	0.54
$p_X(x)$	0.29	0.47	0.24	1.00

The random variables X and Y are not independent since, for example,

$$0.20 = \text{Prob}\{X=0, Y=0\} \neq \text{Prob}\{X=0\}\text{Prob}\{Y=0\} = 0.47 \times 0.46.$$

Exercise 4.3.1 Construct the joint cumulative probability distribution function of two discrete random variables X and Y from their individual cumulative distributions given below, under the assumption that X and Y are independent.

	$x<0$	$0 \le x < 1$	$1 \le x < 2$	$2 \le x < 3$	$3 \le x < 4$	$x \ge 4$
$F_X(x)$	0.0	0.2	0.5	0.5	0.6	1.0

	$y<0$	$0 \le y < 1$	$1 \le y < 2$	$2 \le y < 3$	$3 \le y < 4$	$y \ge 4$
$F_Y(y)$	0.0	0.3	0.5	0.6	0.7	1.0

Answer 4.3.1

$y \geq 4$	0.00	0.20	0.50	0.50	0.60	1.00
$3 \leq y < 4$	0.00	0.14	0.35	0.35	0.42	0.70
$2 \leq y < 3$	0.00	0.12	0.30	0.30	0.36	0.60
$1 \leq y < 2$	0.00	0.10	0.25	0.25	0.3	0.50
$0 \leq y < 1$	0.00	0.06	0.15	0.15	0.18	0.30
$y < 0.00$	0.00	0.00	0.00	0.00	0.00	0.00
$F_{X,Y}(x,y)$	$x < 0$	$0 \leq x < 1$	$1 \leq x < 2$	$2 \leq x < 3$	$3 \leq x < 4$	$x \geq 4$

Exercise 4.4.1 The joint probability density function of two continuous random variables is given by

$$f_{X,Y}(x,y) = \alpha\, xy.$$

What is the value of α if the regions of positive probability are

(a) $0 \leq x \leq 1, \;\; 0 \leq y \leq 1$?

(b) $0 \leq x < y \leq 1$?

Answer 4.4.1 (a) Since

$$\int_0^1 \int_0^1 \alpha\, xy\, dx\, dy = \int_0^1 \alpha y \left[\int_0^1 x\, dx \right] dy = \int_0^1 \alpha y \left[\frac{x^2}{2} \bigg|_0^1 \right] dy = \frac{\alpha}{2} \int_0^1 y\, dy$$

$= \alpha/4$ and it follows that $\alpha = 4$.

(b)

$$\begin{aligned} 1 = \int_0^1 \int_0^y \alpha\, xy\, dx\, dy &= \alpha \int_0^1 y \left[\int_0^y x\, dx \right] dy = \alpha \int_0^1 y \left[\frac{x^2}{2} \bigg|_0^y \right] dy \\ &= \frac{\alpha}{2} \int_0^1 y^3\, dy = \frac{\alpha}{2} \left[\frac{y^4}{4} \bigg|_0^1 \right] = \frac{\alpha}{8}, \end{aligned}$$

and hence $\alpha = 8$.

Exercise 4.4.2 X and Y are two independent and identically distributed continuous random variables. The probability density function of X is

$$f_X(x) = \begin{cases} 3x^2, & 0 \leq x \leq 1, \\ 0 & \text{otherwise}. \end{cases}$$

That of Y is similarly defined. Write down the joint probability density function of X and Y and find the probability that $\text{Prob}\{Y - X \geq 1/2\}$.

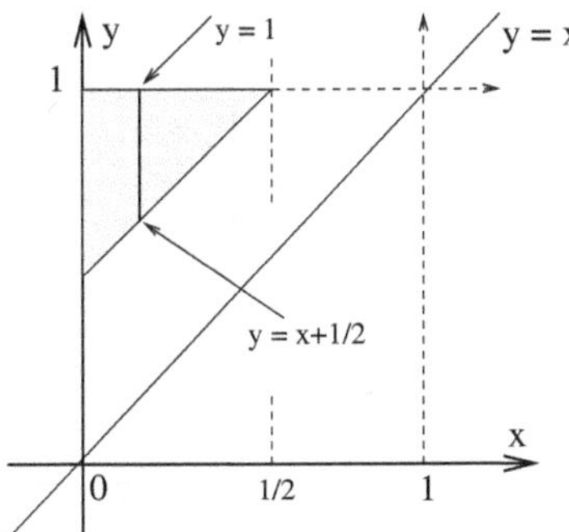

Figure 4.1: Region of positive probability and integration limits for Exercise 4.4.2.

Answer 4.4.2 Since the two random variables are independent, the joint probability density function of X and Y is

$$f_{X,Y}(x,y) = \begin{cases} 9x^2y^2, & 0 \le x \le 1,\ 0 \le y \le 1, \\ 0 & \text{otherwise.} \end{cases}$$

The requested probability is found as

$$\text{Prob}\{Y - X \ge 0.5\} = \text{Prob}\{X \le Y - 0.5\} = \int_{x=0}^{1/2} \int_{y=x+1/2}^{1} 9x^2y^2 \, dy \, dx$$

$$\begin{aligned} &= \int_{x=0}^{1/2} (3x^2y^3) \,|_{y=x+1/2}^{1} \, dx = \int_{x=0}^{1/2} (3x^2 - 3x^2(x+1/2)^3) \, dx \\ &= \int_{x=0}^{1/2} [3x^2 - 3x^2(x^3 + 3x^2/2 + 3x/4 + 1/8)] \, dx \\ &= \int_{x=0}^{1/2} [3x^2 - 3x^5 - 9x^4/2 - 9x^3/4 - 3x^2/8] \, dx \\ &= (x^3 - x^6/2 - 9x^5/10 - 9x^4/16 - x^3/8) \,|_{x=0}^{1/2} = 0.03828. \end{aligned}$$

Exercise 4.4.3 The roots of a quadratic polynomial $ax^2 + bx + c = 0$ are real when $b^2 - 4ac \ge 0$. Use this result to find the probability that the roots of $r^2 + 2Xr + Y = 0$ are real, where X and Y are independent, continuous random variables with X uniformly distributed over $(-1, 1)$ and Y and uniformly distributed over $(0, 1)$.

Answer 4.4.3 In essence, this question requires us to find $\text{Prob}\{X^2 \ge Y\}$. Since X and Y are independent and uniformly distributed, the first over $(-1, 1)$ and the second

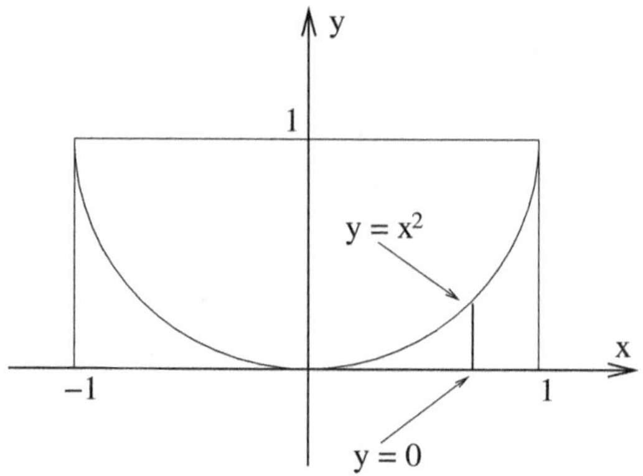

Figure 4.2: Region of positive probability and integration limits for Exercise 4.4.3.

over $(0,1)$, we have

$$\begin{aligned}\text{Prob}\{X^2 \geq Y\} &= \int_{x=-1}^{1}\int_{y=0}^{x^2} f_X(x)f_Y(y)\; dydx = \int_{x=-1}^{1}\int_{y=0}^{x^2} \frac{1}{2}\; dydx \\ &= \int_{-1}^{1} \frac{y}{2}\Big|_0^{x^2} dx = \int_{-1}^{1}\frac{x^2}{2}dx = x^3/6\;\Big|_{-1}^{1} = 1/3.\end{aligned}$$

Exercise 4.4.4 Let X and Y be two continuous random variables whose joint density function is given by

$$f_{X,Y}(x,y) = \begin{cases} \alpha e^{-y}, & 0 < x < y < \infty, \\ 0 & \text{otherwise.}\end{cases}$$

Compute the value of α and then find $F_{X,Y}(2,y)$.

Answer 4.4.4 Observe that

$$\begin{aligned}\int_{-\infty}^{\infty}\int_{-\infty}^{\infty} f_{X,Y}(x,y)dxdy &= \int_0^{\infty}\int_0^{y} \alpha e^{-y}dxdy = \int_0^{\infty} \alpha y e^{-y}dy \\ &= -\alpha y e^{-y}\Big|_0^{\infty} + \int_0^{\infty} \alpha e^{-y}dy = -\alpha e^{-y}\Big|_0^{\infty} = \alpha.\end{aligned}$$

It must follow that $\alpha = 1$. Let us now form $F_{X,Y}(x,y)$ when $x = 2$. Since the density function is only defined for $0 \leq x \leq y \leq \infty$, and our concern is for $X \leq 2$, the area over which we must integrate is bound by $0 \leq x \leq 2$ and $y > x$. We obtain

$$F_{X,Y}(2,y) = \int_0^2 \left(\int_x^y e^{-u}du\right) dx = \int_0^2 \left(e^{-x} - e^{-y}\right) dx = 1 - e^{-2} - 2e^{-y}.$$

Exercise 4.4.5 Let X and Y be two continuous random variables whose joint probability density function is given by

$$f_{X,Y}(x,y) = \begin{cases} 1-\alpha(x+y), & 0 \le x \le 1,\ 0 \le y \le 2, \\ 0 & \text{otherwise.} \end{cases}$$

Find the value of the constant α and the probabilities $\text{Prob}\{X \ge 1/2, Y \le 1\}$ and $\text{Prob}\{X < Y\}$.

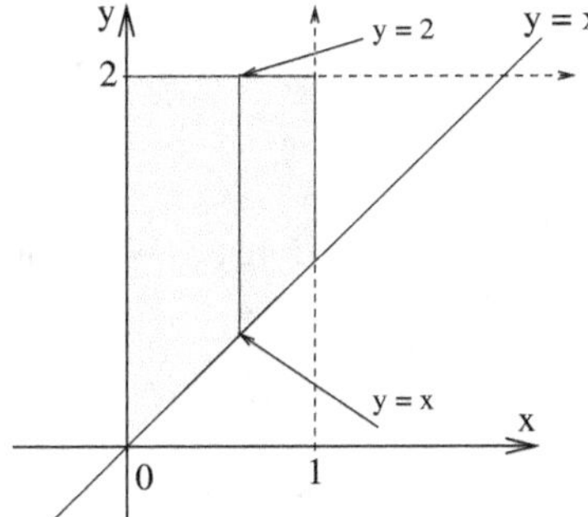

Figure 4.3: Region of positive probability and integration limits for Exercise 4.4.5.

Answer 4.4.5

$$\begin{aligned}
\int_{-\infty}^{\infty}\int_{-\infty}^{\infty} f_{X,Y}(x,y) &= \int_{y=0}^{2}\int_{x=0}^{1} [1-\alpha(x+y)]\, dx\, dy \\
&= \int_{y=0}^{2} \left(x - \alpha x^2/2 - \alpha x y\right)\Big|_{x=0}^{1} dy \\
&= \int_{y=0}^{2} (1-\alpha/2-\alpha y)\, dy \\
&= \left(y - \alpha y/2 - \alpha y^2/2\right)\Big|_{y=0}^{2} = 2 - 3\alpha,
\end{aligned}$$

which yields the result that $\alpha = 1/3$. We now find $\text{Prob}\{X \ge 1/2, Y \le 1\}$:

$$\begin{aligned}
\text{Prob}\{X \ge 1/2, Y \le 1\} &= \int_{x=1/2}^{1}\int_{y=0}^{1} [1-(x+y)/3]\, dy dx \\
&= \int_{x=1/2}^{1} \left(y - xy/3 - y^2/6\right)\Big|_{y=0}^{1} dx \\
&= \int_{x=1/2}^{1} (5/6 - x/3)\, dx \\
&= \left(5x/6 - x^2/6\right)\Big|_{x=1/2}^{1} = 7/24.
\end{aligned}$$

To compute Prob$\{X < Y\}$, we use

$$\begin{aligned}
\text{Prob}\{X < Y\} &= \int_{x=0}^{1}\int_{y=x}^{2} [1-(x+y)/3]\, dydx \\
&= \int_{x=0}^{1} (y - xy/3 - y^2/6)\Big|_{y=x}^{2} dx \\
&= \int_{x=0}^{1} (x^2/2 - 5x/3 + 4/3)dx \\
&= (x^3/6 - 5x^2/6 + 4x/3)\Big|_{x=0}^{1} = 2/3.
\end{aligned}$$

Exercise 4.5.1 The joint probability density function of two random variables X and Y is given by

$$f_{X,Y}(x,y) = \begin{cases} 2, & 0 \le y \le x \le 1, \\ 0 & \text{otherwise.} \end{cases}$$

Derive the conditional probability density functions $f_{X|Y}(x|y)$ and $f_{Y|X}(y|x)$.

Answer 4.5.1 We first compute the marginal density functions. We obtain

$$\begin{aligned}
f_X(x) &= \int_{-\infty}^{\infty} f_{X,Y}(x,y)dy = \int_{0}^{y=x} 2\, dy = 2y\ \big|_0^{y=x} = 2x, \quad 0 \le x \le 1 \\
f_Y(y) &= \int_{-\infty}^{\infty} f_{X,Y}(x,y)dx = \int_{x=y}^{1} 2\, dx = 2x\ \big|_{x=y}^{1} = 2(1-y), \quad 0 \le y \le 1
\end{aligned}$$

We can now find the conditional distributions. We have

$$\begin{aligned}
f_{X|Y}(x|y) &= \frac{f_{X,Y}(x,y)}{f_Y(y)} = \frac{1}{1-y}, \quad y \le x \le 1, \quad \text{and 0 otherwise,} \\
f_{Y|X}(y|x) &= \frac{f_{X,Y}(x,y)}{f_X(x)} = \frac{1}{x}, \quad 0 \le y \le x, \quad \text{and 0 otherwise.}
\end{aligned}$$

Exercise 4.5.2 Let X and Y be two continuous random variables whose joint probability density function is given by

$$f_{X,Y}(x,y) = \begin{cases} 1-(x+y)/3, & 0 \le x \le 1,\ 0 \le y \le 2, \\ 0 & \text{otherwise.} \end{cases}$$

What are the marginal density functions of X and Y? Compute the probability density function of Y given X and hence the following probabilities: (a) Prob$\{Y \le 1 \,|\, X = 0.5\}$ and (b) Prob$\{Y \le 1 \,|\, X = 0.25\}$.

Answer 4.5.2 The marginal density functions of X and Y are

$$\begin{aligned} f_X(x) = \int_{-\infty}^{\infty} f_{X,Y}(x,y)dy &= \int_{y=0}^{2} [1-(x+y)/3]\ dy = (y - xy/3 - y^2/6)\big|_0^2 \\ &= 4/3 - 2x/3, \quad 0 \le x \le 1. \end{aligned}$$

and

$$\begin{aligned} f_Y(y) = \int_{-\infty}^{\infty} f_{X,Y}(x,y)dx &= \int_{x=0}^{1} [1-(x+y)/3]\ dx = (x - xy/3 - x^2/6)\big|_0^1 \\ &= 5/6 - y/3, \quad 0 \le y \le 2. \end{aligned}$$

To compute the probabilities, we first find the conditional density function. From

$$f_{Y|X}(y|x) = \frac{f_{X,Y}(x,y)}{f_X(x)},$$

we obtain

$$f_{Y|X}(y|0.5) = \frac{f_{X,Y}(0.5,y)}{f_X(0.5)} = \frac{1-1/6-y/3}{4/3-1/3} = 5/6 - y/3.$$

It follows that

$$\begin{aligned} \text{Prob}\{Y \le 1 \,|\, X = 0.5\} &= \int_{y \le 1} f_{Y|X}(y|0.5)\ dy = \int_0^1 (5/6 - y/3)\ dy \\ &= \left(\frac{5y}{6} - \frac{y^2}{6}\right)\bigg|_0^1 = 2/3. \end{aligned}$$

Also,

$$f_{Y|X}(y|0.25) = \frac{f_{X,Y}(0.25,y)}{f_X(0.25)} = \frac{1-1/12-y/3}{4/3-1/6} = 11/14 - 2y/7.$$

It follows that

$$\begin{aligned} \text{Prob}\{Y \le 1 \,|\, X = 0.25\} &= \int_{y \le 1} f_{Y|X}(y|0.25)\ dy = \int_0^1 \left(\frac{11-4y}{14}\right)\ dy \\ &= \left(\frac{11y}{14} - \frac{2y^2}{14}\right)\bigg|_0^1 = 9/14. \end{aligned}$$

54 Joint Random Variables and their Distributions

Chapter 5

Expectations and More

Exercise 5.1.1 A random variable X takes values $0, 1, 2, \ldots, n, \ldots$ with probabilities

$$\frac{1}{2}, \frac{1}{3}, \frac{1}{3^2}, \frac{1}{3^3}, \cdots, \frac{1}{3^n} \cdots .$$

Show that this represents a genuine probability mass function. Find $E[X]$, the expected value of X.

Answer 5.1.1 We must show that the sum of the probabilities is equal to 1. Using the standard formula for a geometric series, we find

$$\sum_{i=1}^{\infty}(1/3)^i = \frac{1/3}{1-1/3} = \frac{1}{2}$$

which when added to the value of the first term shows that the probabilities sum to 1. The expected value is given by

$$0 \times (1/2) + 1 \times (1/3) + 2 \times (1/3)^2 + \cdots \quad = \sum_{i=1}^{\infty} i(1/3)^i.$$

Consider the computation of $\sum_{i=1}^{\infty} i\alpha^i$ for $0 < \alpha < 1$. We have

$$\begin{aligned}\sum_{i=1}^{\infty} i\alpha^i &= \alpha\sum_{i=1}^{\infty} i\alpha^{i-1} = \alpha\sum_{i=1}^{\infty}\frac{\partial}{\partial\alpha}\alpha^i = \alpha\frac{\partial}{\partial\alpha}\sum_{i=1}^{\infty}\alpha^i \\ &= \alpha\frac{\partial}{\partial\alpha}\left(\frac{\alpha}{1-\alpha}\right) = \alpha\left(\frac{1}{(1-\alpha)^2}\right) = \frac{\alpha}{(1-\alpha)^2}.\end{aligned}$$

Substituting the value $\alpha = 1/3$, we obtain $E[X] = 0.75$.

Exercise 5.1.2 Two fair dice are thrown. Let X be the random variable that denotes the number of spots shown on the first die and Y the number of spots that show on the second die. It follows that X and Y are independent and identically distributed. Compute $E[X^2]$ and $E[XY]$ and observe that they are not the same.

Answer 5.1.2 The random variables X and Y assume values $1, 2, \ldots, 6$ and each has expectation

$$E[X] = E[Y] = \frac{1+2+3+4+5+6}{6} = \frac{7}{2}.$$

The random variable X^2 assumes values $1, 4, 9, 16, 25$ and 36 and its expectation is

$$E[X^2] = \frac{1+4+9+16+25+36}{6} = \frac{91}{6},$$

while the expectation of the product XY is

$$E[XY] = E[X]E[Y] = \frac{7}{2} \times \frac{7}{2} = \frac{49}{4}.$$

Exercise 5.1.3 Balls are drawn from an urn containing w white balls and b black balls until a white ball appears. Find the mean value and the variance of the number of balls drawn, assuming that each ball is replaced after being drawn.

Answer 5.1.3 We begin by computing the mean number of black balls drawn before a white ball is drawn. In this interpretation, we do not count the time that the white ball is drawn. The probability that a white ball is drawn from the total of $w + b$ balls is given by $w/(b + w)$ and the probability of choosing a black ball is $b/(b + w)$.
The probability that no black ball is drawn before a white ball is given by $w/(w+b)$; the probability that one black ball is drawn before a white ball is $w/(w+b) \times b/(w+b)$; the probability that two black balls are drawn before a white ball is $w/(w+b) \times [b/(w+b)]^2$, and so on. It follows that the mean number of black balls drawn before a white ball is

$$\begin{aligned} E[X] &= 0 \times \left(\frac{w}{w+b}\right) + 1 \times \left(\frac{w}{w+b}\right)\left(\frac{b}{w+b}\right) \\ &+ 2 \times \left(\frac{w}{w+b}\right)\left(\frac{b}{w+b}\right)^2 + 3 \times \left(\frac{w}{w+b}\right)\left(\frac{b}{w+b}\right)^3 + \cdots \\ &= \left(\frac{w}{w+b}\right)\sum_{i=1}^{\infty} i \left(\frac{b}{w+b}\right)^i = \alpha \sum_{i=1}^{\infty} i\beta^i, \end{aligned}$$

where $\alpha = w/(b + w)$, $\beta = b/(b + w)$ and $\alpha + \beta = 1$. Observe that

$$\begin{aligned} \sum_{i=1}^{\infty} i\beta^i &= \beta \sum_{i=1}^{\infty} i\beta^{i-1} = \beta \sum_{i=1}^{\infty} \frac{\partial}{\partial\beta}\beta^i = \beta\frac{\partial}{\partial\beta}\left(\sum_{i=1}^{\infty}\beta^i\right) \\ &= \beta\frac{\partial}{\partial\beta}\left(\frac{\beta}{1-\beta}\right) = \beta\frac{1}{(1-\beta)^2} = \frac{\beta}{(1-\beta)^2}. \end{aligned}$$

It follows then that

$$E[X] = \alpha\frac{\beta}{\alpha^2} = \frac{\beta}{\alpha} = \frac{b/(w+b)}{w/(w+b)} = \frac{b}{w}.$$

If the white ball that terminates the experiment is also to be counted, the mean value in this case increases by 1, to $1 + b/w$. This may be verified independently by computing the sum

$$E'[X] = \alpha + 2\alpha\beta + 3\alpha\beta^2 + 4\alpha\beta^3 + \cdots$$

$$= \alpha + \alpha\sum_{i=1}^{\infty}(i+1)\beta^i = \alpha + \alpha\sum_{i=1}^{\infty} i\beta^i + \alpha\sum_{i=1}^{\infty}\beta^i = \alpha + E[X] + \alpha\frac{\beta}{1-\beta} = E[X] + 1.$$

The variance must be the same for both interpretations. We shall compute it from the relationship

$$\text{Var}[X] = E[X^2] - \left(E[X]\right)^2.$$

We first find the second moment, $E[X^2]$, from

$$\begin{aligned} E[X^2] &= 0^2 \times \left(\frac{w}{w+b}\right) + 1^2 \times \left(\frac{w}{w+b}\right)\left(\frac{b}{w+b}\right) \\ &+ 2^2 \times \left(\frac{w}{w+b}\right)\left(\frac{b}{w+b}\right)^2 + 3^2 \times \left(\frac{w}{w+b}\right)\left(\frac{b}{w+b}\right)^3 + \cdots \\ &= \frac{w}{w+b}\left[\left(\frac{b}{w+b}\right) + 2^2\left(\frac{b}{w+b}\right)^2 + 3^2\left(\frac{b}{w+b}\right)^3 + \cdots\right] \\ &= \alpha\sum_{i=1}^{\infty} i^2\beta^i. \end{aligned}$$

We now show that

$$\sum_{i=1}^{\infty} i^2\beta^i = \frac{\beta(1+\beta)}{(1-\beta)^3} = \frac{\beta(1+\beta)}{\alpha^3},$$

which will then allow us to write

$$\text{Var}[X] = \alpha\frac{\beta(1+\beta)}{\alpha^3} - \frac{\beta^2}{\alpha^2} = \frac{\beta}{\alpha^2} = \frac{b(b+w)}{w^2}.$$

Since

$$\frac{\partial^2}{\partial\beta^2}\beta^{i+2} = (i+2)(i+1)\beta^i,$$

it follows that

$$i^2\beta^i = \frac{\partial^2}{\partial\beta^2}\beta^{i+2} - 3i\beta^i - 2\beta^i$$

and

$$
\begin{aligned}
\sum_{i=1}^{\infty} i^2\beta^i &= \frac{\partial^2}{\partial\beta^2}\sum_{i=1}^{\infty}\beta^{i+2} - 3\sum_{i=1}^{\infty} i\beta^i - 2\sum_{i=1}^{\infty}\beta^i \\
&= \frac{\partial^2}{\partial\beta^2}\left(\frac{\beta^3}{1-\beta}\right) - 3\frac{\beta}{(1-\beta)^2} - 2\frac{\beta}{1-\beta} \\
&= \frac{\partial}{\partial\beta}\left(\frac{3\beta^2-2\beta^3}{(1-\beta)^2}\right) - 3\frac{\beta}{(1-\beta)^2} - 2\frac{\beta}{1-\beta} \\
&= \frac{(1-\beta)^2(6\beta-6\beta^2)+2(1-\beta)(3\beta^2-2\beta^3)}{(1-\beta)^4} \\
&\quad -\frac{3\beta(1-\beta)}{(1-\beta)^3} - \frac{2\beta(1-\beta)^2}{(1-\beta)^3} \quad = \frac{\beta(1+\beta)}{(1-\beta)^3}.
\end{aligned}
$$

Exercise 5.1.4 The joint probability density function of two random variables X and Y is given by

$$
f_{X,Y}(x,y) = \begin{cases} \sin x \sin y, & 0 \le x \le \pi/2,\ 0 \le y \le \pi/2, \\ 0 & \text{otherwise.} \end{cases}
$$

Find the mean and variance of the random variable X.

Answer 5.1.4 Since

$$
f_X(x) = \int_0^{\pi/2} \sin x \sin y\, dy = \sin x, \quad 0 \le x \le \pi/2,
$$

the probability density function for random variable X is

$$
f_X(x) = \begin{cases} \sin x, & 0 \le x \le \pi/2, \\ 0 & \text{otherwise.} \end{cases}
$$

Its expectation is given by

$$
E[X] = \int_0^{\pi/2} x \sin x\, dx = -x\cos x\Big|_0^{\pi/2} + \int_0^{\pi/2} \cos x dx = 0 + \sin x\Big|_0^{\pi/2} = 1
$$

and its variance by

$$
\begin{aligned}
\text{Var}[X] &= E[X^2] - (E[X])^2 = \int_0^{\pi/2} x^2 \sin x\, dx \ - 1 \\
&= -x^2\cos x\Big|_0^{\pi/2} + 2\int_0^{\pi/2} x\cos x\, dx \ - 1 \\
&= 0 + 2\left[x\sin x\Big|_0^{\pi/2} - \int_0^{\pi/2}\sin x\, dx\right] - 1 \\
&= \pi + 2\cos x\Big|_0^{\pi/2} \ - 1 = \pi - 3.
\end{aligned}
$$

Exercise 5.1.5 A stockbroker is interested in studying the manner in which his clients purchase and sell stock. He has observed that these clients may be divided into two equal groups according to their trading habits. Clients belonging to the first group buy and sell stock very quickly (day traders and their ilk); those in the second group buy stocks and hold on to them for a long period. Let X be the random variable that denotes the length of time that a client holds a given stock. The stockbroker observes that the holding time for clients of the first type has mean m_1 and variance λ; for those of the second type, the corresponding numbers are m_2 and μ, respectively. Find the variance of X in terms of m_1, m_2, λ, and μ.

Answer 5.1.5 Using the relations

$$E[X] = \frac{E[X_1] + E[X_2]}{2}, \quad E[X^2] = \frac{E[X_1^2] + E[X_2^2]}{2},$$

we have

$$E[X] = \frac{m_1 + m_2}{2}.$$

Furthermore

$$E[X_1^2] = \text{Var}[X_1] + (E[X_1])^2 = \lambda^2 + m_1^2,$$

$$\text{Var}[X] = E[X^2] - (E[X])^2 = \frac{\lambda^2 + m_1^2 + \mu^2 + m_2^2}{2} - \left(\frac{m_1 + m_2}{2}\right)^2.$$

Hence

$$\text{Var}[X] = \frac{\lambda^2 + \mu^2}{2} + \frac{(m_1 - m_2)^2}{4}.$$

Exercise 5.2.1 Let X and Y be two discrete random variables with joint probability mass function

	$X = -1$	$X = 0$	$X = 1$
$Y = -1$	0.15	0.05	0.20
$Y = 0$	0.05	0.00	0.05
$Y = 1$	0.10	0.05	0.35

Find the expectation of $Z = XY$ and $W = X + Y$.

Answer 5.2.1

$$E[Z] = (-1)(-1)(0.15) + (1)(-1)(0.20) + (-1)(1)(0.10) + (1)(1)(0.35) = 0.20.$$

$$E[W] = -2(0.15) - 0.05 - 0.05 + 0.05 + 0.05 + 2(0.35) = 0.40.$$

Exercise 5.2.2 Let X and Y be two random variables whose joint probability distribution function is given below.

	$X=1$	$X=2$	$X=3$	$X=4$
$Y=-1$	c	0	0	0
$Y=0$	a	$2a$	$2a$	a
$Y=1$	b	$2b$	$2b$	b

(a) Under what conditions does this table represent a proper joint distribution function for the random variables X and Y?

(b) Compute the expectation of the random variables X and Y.

(c) Compute $E[Y^2]$ and $E[(Y-E[Y])^2]$.

Answer 5.2.2

(a) We must have $a \geq 0$, $b \geq 0$, $c \geq 0$ and $6(a+b)+c=1$.

(b) The distribution function of X is given by

$X=1$	$X=2$	$X=3$	$X=4$
$a+b+c$	$2(a+b)$	$2(a+b)$	$a+b$

For Y, its distribution function is given by

$Y=-1$	c
$Y=0$	$6a$
$Y=1$	$6b$

The expectations (computed from $\sum_i x_i p_i$) are as follows:

$$\begin{aligned} E[X] &= 1\times(a+b+c)+2\times 2(a+b)+3\times 2(a+b)+4\times(a+b) \\ &= 15(a+b)+c \\ E[Y] &= -1\times c+0\times 6a+1\times 6b = 6b-c \end{aligned}$$

(c) We compute $E[Y^2]=\sum_i y_i^2 p_i$ as

$$(-1)^2\times c+0^2\times 6a+1^2\times 6b = 6b+c$$

and we compute $E[(Y-E[Y])^2]=\sum_i (y_i-E[Y])^2 p_i$ as

$$(-1-(6b-c))^2c+(0-(6b-c))^2 6a+(1-(6b-c))^2 6b$$

$$= c(c-6b-1)^2+6a(c-6b)^2+6b(c-6b+1)^2.$$

Exercise 5.2.3 The joint probability density function of two random variables X and Y is given by

$$f_{X,Y}(x,y) = \begin{cases} 2, & 0 \le x \le y \le 1, \\ 0 & \text{otherwise.} \end{cases}$$

Find the following quantities:
(a) $E[X]$ and $E[Y]$, (b) $\text{Var}[X]$ and $\text{Var}[Y]$, (c) $E[XY]$ and $\text{Cov}[X,Y]$, and (d) $E[X+Y]$ and $\text{Var}[X+Y]$.

Answer 5.2.3 We first compute the marginal density functions of X and Y.

$$f_X(x) = \int_{-\infty}^{\infty} 2dy = \int_x^1 2dy = 2y\,|_x^1 = 2(1-x), \qquad 0 \le x \le 1,$$

and is zero otherwise.

$$f_Y(y) = \int_{-\infty}^{\infty} 2dx = \int_0^y 2dx = 2x\,|_0^y = 2y, \qquad 0 \le y \le 1,$$

and is zero otherwise. We may now form the required quantities. We have

(a)

$$\begin{aligned} E[X] &= \int_{-\infty}^{\infty} x f_X(x)dx = \int_0^1 2x(1-x)\,dx = \left(x^2 - \frac{2x^3}{3}\right)\Bigg|_0^1 = 1/3, \\ E[Y] &= \int_{-\infty}^{\infty} y f_Y(y)dy = \int_0^1 2y^2\,dy = 2/3. \end{aligned}$$

(b)

$$\begin{aligned} E[X^2] &= \int_{-\infty}^{\infty} x^2 f_X(x)dx = \int_0^1 2x^2(1-x)\,dx = \left(\frac{2x^3}{3} - \frac{2x^4}{4}\right)\Bigg|_0^1 = 1/6, \\ E[Y^2] &= \int_{-\infty}^{\infty} y^2 f_Y(y)dy = \int_0^1 2y^3\,dy = 1/2. \end{aligned}$$

$$\begin{aligned} \text{Var}[X] &= E[X^2] - E[X]^2 = 1/6 - 1/9 = 1/18, \\ \text{Var}[Y] &= E[Y^2] - E[Y]^2 = 1/2 - 4/9 = 1/18. \end{aligned}$$

(c)

$$\begin{aligned} E[XY] &= \int_{x=0}^1 \int_{y=x}^1 2xy\,dy\,dx = \int_0^1 xy^2\Big|_{y=x}^1\,dx \\ &= \int_0^1 (x - x^3)\,dx = \left(\frac{x^2}{2} - \frac{x^4}{4}\right)\Bigg|_0^1 = 1/4, \\ \text{Cov}[X,Y] &= E[XY] - E[X]E[Y] = 1/36, \end{aligned}$$

(d)

$$E[X+Y] = E[X] + E[Y] = 1/3 + 2/3 = 1,$$

$$\text{Var}[X+Y] = \text{Var}[X] + \text{Var}[Y] + 2\,\text{Cov}[X,Y] = 1/18 + 1/18 + 1/18 = 1/6.$$

Exercise 5.2.4 Let X and Y be two random variables whose joint probability density function is

$$f_{X,Y}(x,y) = \begin{cases} x+y, & 0 < x < 1,\ 0 < y < 1, \\ 0 & \text{otherwise.} \end{cases}$$

Find $E[XY]$, $\text{Cov}[X,Y]$, $\text{Corr}[X,Y]$, $E[X+Y]$, and $\text{Var}[X+Y]$.

Answer 5.2.4 We begin by computing the marginal density function of both X and Y.

$$f_X(x) = \int_{-\infty}^{\infty} (x+y)dy = \int_0^1 (x+y)dy = (xy + y^2/2)|_0^1 = x + 1/2, \quad 0 \le x \le 1,$$

and is zero otherwise.

$$f_Y(y) = \int_{-\infty}^{\infty} (x+y)dx = \int_0^1 (x+y)dx = \left(x^2/2 + xy\right)\Big|_0^1 = y + 1/2, \quad 0 \le y \le 1,$$

and is zero otherwise. Thus, X and Y are identically distributed. We now find the expectation and variance of X and of Y, which must be identical.

$$\begin{aligned} E[Y] = E[X] &= \int_{-\infty}^{\infty} x f_X(x)dx = \int_0^1 \left(x^2 + x/2\right) dx \\ &= \left(\frac{x^3}{3} + \frac{x^2}{4}\right)\Big|_0^1 = 7/12, \\ E[Y^2] = E[X^2] &= \int_{-\infty}^{\infty} x^2 f_X(x)dx = \int_0^1 x^2(x+1/2)\,dx \\ &= \left(\frac{x^4}{4} + \frac{x^3}{6}\right)\Big|_0^1 = 5/12, \\ \text{Var}[Y] = \text{Var}[X] &= E[X^2] - E[X]^2 = 5/12 - 49/144 = 11/144. \end{aligned}$$

The required quantities may now be computed. We have

$$\begin{aligned} E[XY] &= \int_{y=0}^1 \int_{x=0}^1 xy(x+y)\,dx\,dy = \int_0^1 \left(\frac{x^3y}{3} + \frac{x^2y^2}{2}\right)\Big|_{x=0}^1 dy \\ &= \int_0^1 \left(\frac{y}{3} + \frac{y^2}{2}\right) dy = \left(\frac{y^2}{6} + \frac{y^3}{6}\right)\Big|_0^1 = 1/3. \end{aligned}$$

$$\text{Cov}[X,Y] = E[XY] - E[X]E[Y] = 1/3 - 7/12 \times 7/12 = -1/144.$$
$$\text{Corr}[X,Y] = \frac{\text{Cov}[X,Y]}{\sigma_X \sigma_Y} = \frac{-1/144}{\sqrt{11/144}\sqrt{11/144}} = -\frac{1}{11}.$$

$$E[X+Y] = E[X] + E[Y] = 7/6.$$
$$\text{Var}[X+Y] = \text{Var}[X] + \text{Var}[Y] + 2\,\text{Cov}[X,Y] = 11/144 + 11/144 - 2/144 = 5/36.$$

Exercise 5.2.5 Let X, Y, and Z be three discrete random variables for which

$$\begin{aligned} \text{Prob}\{X=1,\ Y=1,\ Z=0\} &= p, \\ \text{Prob}\{X=1,\ Y=0,\ Z=1\} &= (1-p)/2, \\ \text{Prob}\{X=0,\ Y=1,\ Z=1\} &= (1-p)/2, \end{aligned}$$

where $0 < p < 1$. Determine the covariance matrix for X and Y.

Answer 5.2.5 In a previous exercise we saw that the joint probability mass distribution is given by

	$X=0$	$X=1$
$Y=0$	0	$(1-p)/2$
$Y=1$	$(1-p)/2$	p

Their covariance matrix is given by

$$\begin{pmatrix} (1-p^2)/4 & -(1-p)^2/4 \\ -(1-p)^2/4 & (1-p^2)/4 \end{pmatrix}.$$

Exercise 5.2.6 The joint density function of two continuous random variables X and Y is given below.

$$f_{X,Y}(x,y) = \begin{cases} 9x^2y^2, & 0 \le x \le 1,\ 0 \le y \le 1, \\ 0 & \text{otherwise.} \end{cases}$$

Find $E[Y|x]$.

Answer 5.2.6 The marginal density function of X is

$$f_X(x) = \begin{cases} 3x^2, & 0 \le x \le 1, \\ 0 & \text{otherwise.} \end{cases}$$

We may now compute $E[Y|x]$ as

$$E[Y|x] = \frac{\int_{-\infty}^{\infty} y f_{X,Y}(x,y)\,dy}{f_X(x)} = \frac{1}{3x^2}\int_0^1 9x^2y^3\,dy = \int_0^1 3y^3\,dy = 3/4,$$

which is independent of x.

Exercise 5.2.7 The joint probability density function of X and Y is given by

$$f_{X,Y}(x,y) = \begin{cases} 1/2, & -1 \le x \le y \le 1, \\ 0 & \text{otherwise.} \end{cases}$$

Find $f_{X|Y}(x|y)$, $f_{Y|X}(x|y)$, $E[X|y]$, and $E[Y|x]$.

Answer 5.2.7 We first compute the marginals of X and Y. We have

$$\begin{aligned} f_X(x) &= \int_{-\infty}^{\infty} f_{X,Y} dy = \int_x^1 \frac{dy}{2} = (1-x)/2, \quad -1 \le x \le 1, \\ f_Y(y) &= \int_{-\infty}^{\infty} f_{X,Y} dx = \int_{-1}^y \frac{dx}{2} = (y+1)/2, \quad -1 \le y \le 1. \end{aligned}$$

Both density function have the value zero outside the indicated limits. We may now compute the conditional probability density functions.

$$\begin{aligned} f_{X|Y}(x|y) &= \frac{f_{X,Y}(x,y)}{f_Y(y)} = \frac{1}{1+y}, \quad -1 \le x \le y, \\ f_{Y|X}(y|x) &= \frac{f_{X,Y}(x,y)}{f_X(x)} = \frac{1}{1-x}, \quad x \le y \le 1. \end{aligned}$$

The conditional expectations are

$$\begin{aligned} E[X|y] &= \int_{-\infty}^{\infty} x f_{X|Y}(x|y)\, dx = \int_{-1}^{y} \frac{x}{1+y} dx \\ &= \left. \frac{x^2}{2(1+y)} \right|_{-1}^{y} = \frac{y^2-1}{2(1+y)} = \frac{y-1}{2}, \\ E[Y|x] &= \int_{-\infty}^{\infty} y f_{Y|X}(y|x)\, dy = \int_x^1 \frac{y}{1-x} dy \\ &= \left. \frac{y^2}{2(1-x)} \right|_x^1 = \frac{1-x^2}{2(1-x)} = \frac{x+1}{2}. \end{aligned}$$

Exercise 5.2.8 Let X and Y be two random variables. Prove the following:

(a) $E[XY] = E[X]E[Y]$, when X and Y are independent.

(b) $E\,[\,E[Y|X]\,] = E[Y]$ for jointly distributed random variables.

(c) $\text{Var}[\alpha X + \beta] = \alpha^2 X$ for constants α and β.

(d) For jointly distributed random variables X and Y, show that the variance of Y is equal to the sum of the expectation of the conditional variance of Y given X and the variance of the conditional expectation of Y given X, i.e., that

$$\text{Var}[Y] = E\left[\text{Var}[Y|X]\right] + \text{Var}\left[E[Y|X]\right].$$

Answer 5.2.8

(a) Let X and Y be independent *discrete* random variables Then

$$E[XY] = \sum_{-\infty}^{\infty}\sum_{-\infty}^{\infty} xyp_X(x)p_Y(y) = \sum_{-\infty}^{\infty} xp_X(x)\sum_{-\infty}^{\infty} yp_Y(y) = E[X]E[Y].$$

For *continuous* random variables, replace the summations with integrals.

(b)

$$\begin{aligned} E\left[\,E[Y|X]\,\right] &= \int_{-\infty}^{\infty} E[Y|x]f_X(x)\,dx = \int_{-\infty}^{\infty}\int_{-\infty}^{\infty} yf_{Y|X}(y|x)f_X(x)\;dy\,dx \\ &= \int_{-\infty}^{\infty} y\int_{-\infty}^{\infty} f_{X,Y}(x,y)\;dx\,dy = \int_{-\infty}^{\infty} yf_Y(y)\,dy = E[Y]. \end{aligned}$$

(c) Let $E[X] = \mu_X$. Then

$$\text{Var}[\alpha X + \beta] = E\left[(\alpha X + \beta - \alpha\mu_X - \beta)^2\right] = E\left[\alpha^2(X - \mu_X)^2\right] = \alpha^2\text{Var}[X].$$

(d) Using the definition of conditional variance and part (b), we proceed as follows.

$$\begin{aligned} E[\,\text{Var}[Y|x]\,] &= E\left[\,E[Y^2|X] - E[Y|X]^2\,\right] \\ &= E[Y^2] - E\left[E[Y|X]^2\right] \\ &= E[Y^2] - \left(\text{Var}\left[E[Y|X]\right] + E\left[E[Y|X]\right]^2\right) \\ &= E[Y^2] - \text{Var}\left[E[Y|X]\right] - E[Y]^2 \\ &= \text{Var}[Y] - \text{Var}[\,E[Y|X]\,]. \end{aligned}$$

Exercise 5.3.1 A discrete random variable X takes the value 1 if the number 6 appears on a single throw of a fair die and takes the value 0 otherwise. Find its probability generating function.

Answer 5.3.1 We have $\text{Prob}\{X = 1\} = p = 1/6$ and $\text{Prob}\{X = 0\} = q = 1 - p = 5/6$. In answering this question, we shall use the probabilities p and q rather than their numeric values 1/6 and 5/6 respectively . The generating function of X is given by

$$G_X(z) = \sum_{i=0}^{\infty} p_i z^i = p_0 z^0 + p_1 z^1 = qz^0 + pz^1 = q + pz = (1 - p) + pz = 1 - p(1 - z).$$

Exercise 5.3.2 The probability mass function of a random variable X is given by

$$p_k = \text{Prob}\{X = k) = p_X(k) = \begin{cases} \binom{n}{k} p^k q^{n-k}, & 0 \le k \le n, \\ 0 & \text{otherwise,} \end{cases}$$

where p and q are nonzero probabilities such that $p + q = 1$. This distribution is referred to as the *binomial distribution* with parameters n and p. Find the probability generating function of X. Also find the probability generating function of the sum of m such random variables, X_i, $i = 1, 2, \ldots, m$, with parameters n_i and p, respectively.

Answer 5.3.2 The probability generating function of X is

$$G_X(z) = \sum_{k=0}^{n} \binom{n}{k} p^k q^{n-k} z^k = \sum_{k=0}^{n} \binom{n}{k} (pz)^k q^{n-k}.$$

Applying the binomial theorem,

$$(a+b)^n = \sum_{k=0}^{n} \binom{n}{k} a^k b^{n-k},$$

we obtain the probability generating function of X as

$$G_X(z) = (pz + q)^n.$$

To answer the second part of the question, let $X = X_1 + X_2 + \cdots + X_m$ with $G_{X_i}(z) = (pz + q)^{n_i}$ for $i = 1, 2, \ldots, m$. Then

$$\begin{aligned} G_X(z) &= G_{X_1}(z) G_{X_2}(z) \cdots G_{X_1}(z) = (pz+q)^{n_1} (pz+q)^{n_2} \cdots (pz+q)^{n_m} \\ &= (pz+q)^{n_1+n_2+\cdots+n_m} \end{aligned}$$

which is the probability generating function of a random variable having a binomial distribution with parameters $n_1 + n_2 + \cdots + n_m$ and p.

Exercise 5.3.3 Consider a discrete random variable X whose probability generating function is given by

$$p_k = p_X(k) = \begin{cases} \alpha^k e^{-\alpha}/k!, & k = 0, 1, 2, \ldots, \quad \alpha > 0, \\ 0 & \text{otherwise.} \end{cases}$$

This distribution is referred to as the *Poisson distribution* with parameter α. Find the probability generating function of X. Also find the probability generating function of the sum of m such random variables, X_i, $i = 1, 2, \ldots, m$, with parameters α_i, respectively.

Answer 5.3.3 The probability generating function of X is

$$G_X(z) = \sum_{k=0}^{n} e^{-\alpha} \frac{\alpha^k z^k}{k!} = e^{-\alpha} \sum_{k=0}^{n} \frac{(\alpha z)^k}{k!} = e^{-\alpha} e^{\alpha z} = e^{\alpha(z-1)}.$$

Now let $X = X_1 + X_2 + \cdots + X_m$ with $G_{X_i}(z) = e^{\alpha_i(z-1)}$ for $i = 1, 2, \ldots, m$. Then

$$\begin{aligned} G_X(z) &= G_{X_1}(z)G_{X_2}(z)\cdots G_{X_1}(z) = e^{\alpha_1(z-1)}e^{\alpha_2(z-1)}\cdots e^{\alpha_m(z-1)} \\ &= e^{(\alpha_1+\alpha_2+\cdots+\alpha_m)(z-1)}. \end{aligned}$$

Exercise 5.4.1 Find the moment generating function of the discrete random variable X with probability mass function given in Problem 5.4.2. Derive the value of $E[X]$ from this moment generating function. In Problem 5.4.2 the probability mass function of X is

$$p_X(x) = \begin{cases} 1/10, & x = 1, \\ 2/10, & x = 2, \\ 3/10, & x = 3, \\ 4/10, & x = 4, \\ 0 & \text{otherwise.} \end{cases}$$

Answer 5.4.1 The probability mass function is given by

$$p_X(x) = \begin{cases} 1/10, & k = 1, \\ 2/10, & k = 2, \\ 3/10, & k = 3, \\ 4/10, & k = 4, \\ 0 & \text{otherwise.} \end{cases}$$

$$\mathcal{M}_X(\theta) = \sum_{k=1}^{\infty} e^{k\theta} p_X(k) = \frac{e^{\theta}}{10} + \frac{2e^{2\theta}}{10} + \frac{3e^{3\theta}}{10} + \frac{4e^{4\theta}}{10}.$$

The expectation of X may be found from the first derivative of $\mathcal{M}_X(\theta)$. We have

$$E[X] = \mathcal{M}'_X(\theta)|_{\theta=0} = \left.\left(\frac{e^{\theta}}{10} + \frac{4e^{2\theta}}{10} + \frac{9e^{3\theta}}{10} + \frac{16e^{4\theta}}{10}\right)\right|_{\theta=0} = 3.$$

Exercise 5.4.2 Consider a discrete random variable X whose probability mass function is given by

$$p_X(k) = \begin{cases} 3 - e, & k = 0, \\ 0, & k = 1, \\ 1/k!, & k = 2, 3, \ldots, \\ 0 & \text{otherwise.} \end{cases}$$

Find the moment generating function of X and from it compute $E[X]$.

Answer 5.4.2

$$\mathcal{M}_X(\theta) = \sum_{k=0}^{\infty} e^{k\theta} p_X(k) = 3 - e + \sum_{k=2}^{\infty} \frac{e^{k\theta}}{k!}.$$

The expectation of X may be found from the first derivative of $\mathcal{M}_X(\theta)$.

$$\begin{aligned} E[X] = \mathcal{M}'_X(\theta)|_{\theta=0} &= \frac{d}{d\theta}\left(3 - e + \frac{e^{2\theta}}{2!} + \frac{e^{3\theta}}{3!} + \frac{e^{4\theta}}{4!} + \cdots\right)\Bigg|_{\theta=0} \\ &= \left(\frac{2e^{2\theta}}{2!} + \frac{3e^{3\theta}}{3!} + \frac{4e^{4\theta}}{4!} + \cdots\right)\Bigg|_{\theta=0} \\ &= 1 + \frac{1}{2!} + \frac{1}{3!} + \cdots = e - 1. \end{aligned}$$

Exercise 5.4.3 The moment generating function of a discrete random variable X is

$$\mathcal{M}_X(\theta) = \frac{e^{\theta}}{12} + \frac{e^{3\theta}}{3} + \frac{e^{6\theta}}{6} + \frac{e^{9\theta}}{3} + \frac{e^{12\theta}}{12}.$$

Find the probability mass function of X.

Answer 5.4.3 By comparing coefficients of $e^{k\theta}$, we obtain

k	1	3	6	9	12	otherwise
$p_X(k)$	1/12	1/3	1/6	1/3	1/12	0

Exercise 5.5.1 Alice and Bob, working with an architect, have drawn up plans for their new home, and are ready to send them out to the best builders in town to get estimates on the cost of building this home. At present they have identified five different builders. Their architect has told them that the estimates should be uniformly distributed between \$500,000 and \$600,000. What is the cumulative distribution and the expectation of the lowest estimate, assuming that all builders, and hence their estimates, are independent? Compute the probability that the lowest estimate does not exceed \$525,000. What would this distribution, expectation, and probability be, if Alice and Bob were able to identify 20 builders? What conclusions could be drawn concerning the size of the lowest estimate and the number of builders?

Answer 5.5.1 To simplify the notation, we shall work in units of 100,000. The cumulative distribution function of a single estimate is given by

$$F_X(x) = \begin{cases} 0, & x \le 5, \\ x - 5, & 5 \le x \le 6, \\ 1, & x \ge 6. \end{cases}$$

The cumulative distribution function of the minimum of 5 such estimates is

$$F_{X_{\min}} = \begin{cases} 0, & x \leq 5, \\ 1 - [1 - (x-5)]^5, & 5 \leq x \leq 6, \\ 1, & x \geq 6. \end{cases}$$

Writing $1 - (x - 5)$ as $6 - x$, the probability density function of $X_{\min}$ is

$$f_{X_{\min}}(x) = \frac{d}{dx} F_{X_{\min}}(x) = 5(6-x)^4, \quad 5 \leq x \leq 6,$$

and is equal to zero otherwise. Its expectation is found as

$$\begin{aligned} E[X_{\min}] &= 5\int_5^6 x(6-x)^4\,dx = 5\left(-x\,\frac{(6-x)^5}{5}\Bigg|_5^6 + \frac{1}{5}\int_5^6 (6-x)^5 dx\right) \\ &= 5\left(1 - \frac{1}{5}\,\frac{(6-x)^6}{6}\Bigg|_5^6\right) = 5 + \frac{1}{6}, \end{aligned}$$

which corresponds to \$516,667. The probability that the minimum estimate is less than \$525,000 is

$$\text{Prob}\{X \leq 5.25\} = F_{X_{min}}(5.25) = 1 - (0.75)^5 = 0.7627.$$

With 20 estimates, we find

$$F_{X_{\min}} = \begin{cases} 0, & x \leq 5, \\ 1 - (6-x)^{20}, & 5 \leq x \leq 6, \\ 1, & x \geq 6. \end{cases}$$

$$f_{X_{\min}}(x) = \frac{d}{dx} F_{X_{\min}}(x) = 20(6-x)^{19}, \quad 5 \leq x \leq 6,$$

and

$$\begin{aligned} E[X_{\min}] &= 20\int_5^6 x(6-x)^{19}\,dx = 20\left(-x\,\frac{(6-x)^{20}}{20}\Bigg|_5^6 + \frac{1}{20}\int_5^6 (6-x)^{20} dx\right) \\ &= 20 \times \frac{1}{4} + \int_5^6 (6-x)^{20} dx = 5 + \frac{(6-x)^{21}}{21}\Bigg|_5^6 = 5 + \frac{1}{21} = 5.04762, \end{aligned}$$

which corresponds to \$504,762. The probability that the minimum estimate is now less than \$525,000 is

$$\text{Prob}\{X \leq 5.25\} = F_{X_{min}}(5.25) = 1 - (0.75)^{20} = 0.9968.$$

Indeed with probability 0.88, the minimum estimate will be less than \$510,000.
It is apparent that increasing the number of builders who provide estimates reduces the size of the minimum estimate.

Chapter 6

Discrete Probability Distributions

Exercise 6.3.1 The probability of winning a lottery is 0.0002. What is the probability of winning at least twice in 1,000 tries?

Answer 6.3.1

$$\text{Prob}\{k \geq 2\} = 1 - \text{Prob}\{k = 0\} - \text{Prob}\{k = 1\}$$

$$= 1 - \binom{1000}{0}(0.0002)^0(0.9998)^{1000} - \binom{1000}{1}(0.0002)^1(0.9998)^{999}$$

$$= 1 - (0.9998)^{1000} - 0.2(0.9998)^{999} = 0.0175.$$

Exercise 6.3.2 A scientific experiment is carried out a number of times in the hope that a later analysis of the data finds at least one success. Let n be the number of times that the experiment is conducted and suppose that the probability of success is $p = 0.2$. Assuming that the experiments are conducted independently from one another, what is the number of experiments that must be conducted to be 95% sure of having at least one success?

Answer 6.3.2 Since each of the n experiments can be considered an independent Bernoulli trial with probability of success p, the probability of obtaining k successes in n experiments may be obtained from the binomial distribution. In particular, the probability of having at least one success is given by

$$\text{Prob}\{k \geq 1\} = 1 - \text{Prob}\{k = 0\} = 1 - p^0(1-p)^n = 1 - (1-p)^n = 1 - (0.8)^n.$$

The number of experiments that must be conducted to be 95% sure of at least one success is found be setting $1 - (0.8)^n = 0.95$, i.e., $0.8^n = 0.05$. Solving for n, we find

$$n = \frac{\ln 0.05}{\ln 0.8} = 13.4251.$$

Thus, at least 14 experiments need to be conducted to have a 95% chance of seeing at least one success.

Exercise 6.4.1 Derive the expectation and variance of a geometric random variable directly from its definition by associating $n^j q^n$ with its jth derivative with respect to q and interchanging the summation and the derivative.

Answer 6.4.1 In computing the expectation, we use the fact that $nq^{n-1} = d(q^n)/dq$ and proceed as follows:

$$\begin{aligned} E[X] &= \sum_{n=1}^{\infty} npq^{n-1} = p\sum_{n=1}^{\infty} nq^{n-1} = p\sum_{n=0}^{\infty} \frac{d}{dq}q^n \\ &= p\frac{d}{dq}\sum_{n=0}^{\infty} q^n = p\frac{d}{dq}\frac{1}{1-q} = \frac{p}{(1-q)^2} = \frac{1}{p}. \end{aligned}$$

The variance is computed in a similar manner. This time using $n(n-1)q^{n-2} = d^2(q^n)/dq^2$, we find

$$\begin{aligned} E[X^2] &= \sum_{n=1}^{\infty} n^2pq^{n-1} = \sum_{n=1}^{\infty} n(n-1)pq^{n-1} + \sum_{n=1}^{\infty} npq^{n-1} \\ &= q\sum_{n=1}^{\infty} n(n-1)pq^{n-2} + \sum_{n=1}^{\infty} npq^{n-1} = pq\frac{d^2}{dq^2}\sum_{n=1}^{\infty} q^n + p\frac{d}{dq}\sum_{n=1}^{\infty} q^n \\ &= pq\frac{d^2}{dq^2}\frac{q}{1-q} + p\frac{d}{dq}\frac{q}{1-q} = pq\frac{2}{(1-q)^3} + p\frac{1}{(1-q)^2} \\ &= \frac{2pq}{p^3} + \frac{p}{p^2} = \frac{2q+p}{p^2} = \frac{1+q}{p^2}. \end{aligned}$$

Hence

$$\text{Var}[X] = E[X^2] - E[X]^2 = \frac{1+q}{p^2} - \frac{1}{p^2} = \frac{q}{p^2} = \frac{1-p}{p^2}.$$

Exercise 6.4.2 A young couple decides to have children until their first son is born. Assuming that each child born is equally likely to be a boy or a girl, what is the probability that this couple will have exactly four children? What is the most probable range for the number of children this couple will have?

Answer 6.4.2 Let X be the (geometric) random variable that denotes the number of children the couple will have. From the geometric distribution, with parameter $p = 1/2$, the probability of the couple having 3 girls and then a boy is given by

$$p_X(4) = (1-p)^3 p = \left(\frac{1}{2}\right)^4 = \frac{1}{16}.$$

The expected number of children the couple will have is $E[X] = 1/p = 2$ and the standard deviation is $\sigma_X = \sqrt{(1-p)/p^2} = \sqrt{2}$. The likely (2-sigma range) is given by $2 \pm 2\sqrt{2} = [-0.83, 4.83]$. Therefore, in their search for a male heir, the couple is likely to have between 1 and 5 children.

Exercise 6.5.1 Let X be an integer valued random variable whose probability mass function is given by

$$\text{Prob}\{X = n\} = \alpha t^n \quad \text{for all } n \geq 0,$$

where $0 < t < 1$. Find $G_X(z)$, the value of α, the expectation of X, and its variance.

Answer 6.5.1

$$G_X(z) = \alpha \sum_{n=0}^{\infty} (tz)^n = \frac{\alpha}{1 - tz}.$$

We find the constant α from

$$1 = G_X(1) = \frac{\alpha}{1-t}$$

and hence $\alpha = 1-t$. To compute the mean and variance we take the first two derivatives of $G_X(z)$.

$$G'_X(z) = \frac{\alpha t}{(1-tz)^2}, \quad G''_X(z) = \frac{2\alpha t^2}{(1-tz)^3}.$$

Hence

$$E[X] = G'_X(1) = \frac{t}{1-t}$$

and

$$\text{Var}[X] = G''_X(1) + G'_X(1) - (G'_X(1))^2 = \frac{2t^2 + t(1-t) - t^2}{(1-t)^2} = \frac{t}{(1-t)^2}.$$

Observe that the random variable X of this problem is a straightforward modified geometric random variable whose parameter t is equal to the parameter $1-p$ of our original definition.

Exercise 6.5.2 The boulevards leading into Paris appear to drivers as an endless string of traffic lights, one after the other, extending as far as the eye can see. Fortunately, these successive traffic lights are synchronized. Assume that the effect of the synchronization is that a driver has a 95% chance of not being stopped at any light, independent of the number of green lights she has already passed. Let X be the random variable that counts the number of green lights she passes before being stopped by a red light.

(a) What is the probability distribution of X?
(b) How many green lights does she pass on average before having to stop at a red light?
(c) What is the probability that she will get through 20 lights before having to stop for the first time?
(d) What is the probability that she will get through 50 lights before stopping for the fourth time.?

Answer 6.5.2

(a) The random variable X has a modified geometric distribution with parameter $p = 0.05$: the probability of success here is interpreted as the probability of having to stop at a traffic light, and we need to count the number of "failures" (getting through a traffic light without stopping) before the first "success". The probability mass function of X is

$$\text{Prob}\{X = n\} = \begin{cases} p(1-p)^n, & n = 1, 2, \ldots, \\ 0 & \text{otherwise.} \end{cases}$$

(b) The expected number of "failures" before the "success" of having to stop is found as

$$E[X] = \frac{1-p}{p} = \frac{.95}{0.05} = 19.$$

(c)

$$\text{Prob}\{X = 20\} = p(1-p)^{20} = 0.05 \times 0.95^{20} = 0.0179.$$

(d) The probability of getting through 50 sets of traffic lights before stopping for the 4th time may be obtained from the negative binomial distribution with $n = 51$ and $k = 4$, which is

$$\begin{pmatrix} n-1 \\ k-1 \end{pmatrix} p^k (1-p)^{n-k} = \begin{pmatrix} 50 \\ 3 \end{pmatrix} 0.05^4 \times 0.95^{46} = 0.0116.$$

Exercise 6.6.1 A popular morning radio show offers free entrance tickets to the local boat show to the sixth caller who rings the station with the correct answer to a question. Assume that all calls are independent and have probability $p = 0.7$ of being correct. Let X be the random variable that counts the number of calls needed to find the winner.

(a) What is the probability mass function of X?
(b) What is the probability of finding a winner on the 12th call?
(c) What is the probability that it will take more than ten calls to find a winner?

Answer 6.6.1 The random variable X is a negative binomial random variable with parameters $p = 0.7$ and $k = 6$. Consequently,

(a)

$$p_X(n) = \begin{pmatrix} n-1 \\ 6-1 \end{pmatrix} (0.7)^6 (0.3)^{n-6}, \quad n = 6, 7, \ldots,$$

and is equal to zero otherwise.

(b)

$$p_X(12) = \begin{pmatrix} 11 \\ 5 \end{pmatrix} (0.7)^6 (0.3)^6 = 0.396.$$

(c)

$$\text{Prob}\{X > 10\} = 1 - \text{Prob}\{X \leq 10\}$$

$$\begin{aligned} &= 1 - [p_X(6) + p_X(7) + p_X(8) + p_X(9) + p_X(10)] \\ &= 1 - (0.7)^6 \left[1 + 6(0.3) + 21(0.3)^2 + 56(0.3)^3 + 126(0.3)^4\right] \\ &= 1 - (0.7)^6 \, [7.2226] = 0.1503. \end{aligned}$$

Exercise 6.6.2 In a best of seven sports series, the first person to win four games is declared the overall winner. One of the players has a 55% chance of winning each game, independent of any other game. What is the probability that this player wins the series?

Answer 6.6.2 This player may win in $n = 4, 5, 6$ or 7 games. Let X be the (negative binomial) random variable that denotes the number of games up to and including the 4th win. The probability that this player wins the series is then

$$\sum_{n=4}^{7} \text{Prob}\{X = n\} = \sum_{n=4}^{7} \binom{n-1}{4-1} (0.55)^4 \, (0.45)^{n-4}$$

$$= \binom{3}{3} (0.55)^4 \, (0.45)^0 + \binom{4}{3} (0.55)^4 \, (0.45)^1 + \binom{5}{3} (0.55)^4 \, (0.45)^2$$

$$+ \binom{6}{3} (0.55)^4 \, (0.45)^3$$

$$= (0.55)^4 \left[1 + 4 \times 0.45 + 10 \times 0.45^2 + 20 \times 0.45^3\right] = 0.6083.$$

Exercise 6.6.3 Returning to Exercise 6.6.2, find the probability that the series ends after game 5. Once again, assume that one of the players has a 55% change of winning each game, independent of any other game.

Answer 6.6.3 Since either player may be the first to win four games, we need to compute the probability of each of them being declared the winner after game 5. The parameters for the negative binomial distribution are $n = 5$ and $k = 4$, Making the appropriate substitutions, the probability that the first player (with probability 0.55 of winning each game) wins the series in 5 games is

$$\binom{5-1}{4-1} (0.55)^4 \, (0.45)^1 = 0.1647.$$

This is the probability that the first player will obtain his 4th success in game 5. The corresponding numbers for the second player are

$$\binom{4}{3}(0.45)^4\,(0.55)^1 = 0.0902.$$

The probability that the series lasts exactly 5 games is the sum of these, i.e., 0.2549.

Exercise 6.6.4 Given the expansion

$$(1-q)^{-k} = \sum_{i=0}^{\infty}\binom{i+k-1}{k-1}q^i,$$

show that Equation (6.1) defines a bona fide probability mass function. The name *negative* binomial comes from this expansion with its negative exponent, $-k$. Recall that Equation (6.1) of the text is given as

$$p_X(n) = \binom{n-1}{k-1}p^k(1-p)^{n-k}, \qquad k \geq 1, \quad n = k, k+1, \ldots. \tag{6.1}$$

Answer 6.6.4 The probability mass function of a negative binomial random variable, X, is

$$p_X(n) = \binom{n-1}{k-1}p^k(1-p)^{n-k}, \qquad k \geq 1, \quad n = k, k+1, \ldots,$$

which is nonnegative for all defined values. We need to show that the sum over all n is equal to 1. Setting $q = 1-p$ and introducing the change of variable $i = n-k$, we have

$$\sum_{n=k}^{\infty}\binom{n-1}{k-1}p^k q^{n-k} = p^k\sum_{i=0}^{\infty}\binom{i+k-1}{k-1}q^i = p^k\,(1-q)^{-k} = 1.$$

Exercise 6.6.5 A different couple from those of Exercise 6.4.2 intend to continue having children until they have two boys. What is the probability they will have exactly two children? exactly three children? exactly four children? What is the most probable range for the number of children this couple will have?

Answer 6.6.5 Let X be the (negative binomial) random variable that denotes the number of children the couple will have. Then

$$\text{Prob}\{X = n\} = \binom{n-1}{2-1}p^2(1-p)^{n-2}.$$

Therefore,

$$\begin{aligned}\text{Prob}\{X=2\} &= \binom{1}{1}(0.5)^2 = \frac{1}{4},\\ \text{Prob}\{X=3\} &= \binom{2}{1}(0.5)^3 = \frac{1}{4},\\ \text{Prob}\{X=4\} &= \binom{3}{1}(0.5)^4 = \frac{3}{16}.\end{aligned}$$

The expectation and standard deviation of X are

$$E[X] = \frac{k}{p} = \frac{2}{0.5} = 4, \qquad \sigma_X = \frac{\sqrt{k(1-p)}}{p} = \frac{1}{1/2} = 2.$$

The 2-sigma range about the expectation is $4 \pm 4 = [0, 8]$. Since the couple will have at least two children, the most probable range should be taken as $[2, 8]$.

Exercise 6.8.1 I have four pennies and three dimes in my pocket. Using the hypergeometric distribution, compute the probability that, on pulling two coins at random from my pocket, I have enough to purchase a 20-cent newspaper. Compute the answer again, this time using only probabilistic arguments.

Answer 6.8.1 To be able to purchase the newspaper, I need to choose two dimes from the seven coins. Letting X be the number of dimes picked, then X is a hypergeometric random variable with parameters $N = 7$, $r = 3$ and $n = 2$. The probability that two dimes are chosen is

$$\text{Prob}\{X=2\} = \frac{\binom{r}{2}\binom{N-r}{n-2}}{\binom{N}{n}} = \frac{\binom{3}{2}\binom{4}{0}}{\binom{7}{2}} = \frac{3}{21}.$$

The number of ways in which 2 coins can be chosen at random from a collection of 7 coins is C_2^7. The number of ways in which 2 dimes can be chosen from 3 dimes is C_2^3. The probability of choosing two dimes from a set of 3 dimes and 4 pennies, is the ratio of these, i.e.

$$\text{Prob}\{X=2\} = \frac{C_2^3}{C_2^7} = \frac{\binom{3}{2}}{\binom{7}{2}} = \frac{3}{21}.$$

Exercise 6.8.2 Let X be a hypergeometric random variable with parameters $N = 12$, $r = 8$ and $n = 6$.

(a) What are the possible values for X?
(b) What is the probability that X is greater than 2?
(c) Compute the expectation and variance of X.

Answer 6.8.2

(a) The random variable X is bound by the quantities

$$\max(0, n - N + r) = 2, \quad \text{and} \quad \min(n, r) = 6.$$

(b)

$$\text{Prob}\{X > 3\} = 1 - \text{Prob}\{X = 2\} = 1 - \frac{\binom{8}{2}\binom{4}{4}}{\binom{12}{6}} = 1 - \frac{28}{924} = 0.9697.$$

(c) The expectation and variance of X are

$$E[X] = n\left(\frac{r}{N}\right) = 6 \times \frac{8}{12} = 4,$$

$$\text{Var}[X] = E[X]\left(1 - \frac{r}{N}\right)\left(\frac{N-n}{N-1}\right) = 4 \times \frac{4}{12} \times \frac{6}{11} = 0.7273.$$

Exercise 6.8.3 Cardinal Gibbons High School has 1,200 students, 280 of whom are seniors. In a random sample of ten students, compute the expectation and variance of the number of included seniors using both the hypergeometric distribution and its binomial approximation.

Answer 6.8.3 Let X be the number of seniors in the selected group of 10 students. The expectation and variance of X using the hypergeometric distribution with parameters $N = 1,200$, $r = 280$ and $n = 10$, are

$$E[X] = n\left(\frac{r}{N}\right) = 10 \times \frac{280}{1200} = \frac{7}{3},$$

$$\text{Var}[X] = E[X]\left(1 - \frac{r}{N}\right)\left(\frac{N-n}{N-1}\right) = 10 \times \frac{280}{1200} \times \frac{920}{1200} \times \frac{1190}{1199} = 1.7755.$$

Using the binomial distribution with parameter $p = 280/1200$ as an approximation, we obtain

$$E[X^\dagger] = np = 10 \times \frac{280}{1200} = \frac{7}{3},$$

exactly as before, (since we choose $p = r/N$), while for the variance, we obtain

$$\text{Var}[X^\dagger] = np(1-p) = 10 \times \frac{280}{1200} \times \frac{920}{1200} = 1.7889.$$

Exercise 6.9.1 Write down the probability generating function for a Poisson random variable X, and use it to compute its expectation and variance.

Answer 6.9.1 The probability generating function for a discrete Poisson random variable X is given by

$$G_X(z) = E[z^X] = \sum_k z^k p_k = e^{\alpha(z-1)}, \quad \text{for } |z| \le 1.$$

Also, since

$$G'_X(1) = E[X] \quad \text{and} \quad G''_X(1) = E[X(X-1)] = E[X^2] - E[X],$$

it follows that

$$E[X] = \alpha\, e^{\alpha(z-1)}\Big|_{z=1} = \alpha$$

and since

$$\text{Var}[X] = E[X^2] - E[X]^2 = G''_X(1) + G'_X(1) - [G'_X(1)]^2$$

we may conclude that

$$\text{Var}[X] = (\alpha)^2\, e^{\alpha(z-1)}\Big|_{z=1} + \alpha - \alpha^2 = \alpha.$$

Exercise 6.9.2 Consider a situation in which certain events occur randomly in time, such as arrivals to a queueing system. Let $X(t)$ be the number of these events that occur during a time interval of length t. We wish to compute the distribution of the random variable $X(t)$, under the following three assumptions:

(1) The events are independent of each other. This means that $X(\Delta t_1)$, $X(\Delta t_2), \ldots$ are independent if the intervals Δt_1, Δt_2, ... do not overlap.

(2) The system is *stationary*, i.e., the distribution of $X(\Delta t)$ depends only on the length of Δt and not on the actual time of occurrence.

(3) We have the following probabilities:

- Prob{at least 1 event in Δt} $= \lambda \Delta t + o(\Delta t)$,
- Prob{more than 1 event in Δt} $= o(\Delta t)$,

where $o(\Delta t)$ is a quantity that goes to zero faster than Δt, i.e.,

$$\lim_{\Delta t \to 0} \frac{o(\Delta t)}{\Delta t} = 0,$$

and λ is a positive real number that denotes the rate of occurrence of the events. Show that X is a Poisson random variable.

Answer 6.9.2 Consider a time interval $[0, t]$ and define $X(t)$ to be the number of events occurring in this interval. Now divide this interval into n equal time intervals Δt_1, $\Delta t_2, \ldots$ and let $X(\Delta t_k)$ be the random variable that counts the number of events that occur in the interval Δt_k. Then we must have

$$X(t) = \sum_{k=1}^{n} X(\Delta t_k)$$

since the random variables $X(\Delta t_k)$ are independent. The generating function for each of these n random variables, $X(\Delta t)$, is given by

$$G_{X(\Delta t)}(z) = \left(1 - \lambda \frac{t}{n}\right) + \lambda \frac{t}{n} z + o\left(\frac{t}{n}\right).$$

Thus, the generating function of $X(t)$ is

$$G_X(z) = \left[G_{X(\Delta t)}(z)\right]^n = \left[1 + \frac{\lambda t(z-1)}{n} + o\left(\frac{t}{n}\right)\right]^n.$$

Since $G_X(z)$ is independent of the subintervals Δt_1, $\Delta t_2, \ldots$, we may take the limit as $n \to \infty$ and get

$$F(z) = \lim_{n \to \infty} \left[1 + \frac{\lambda t(z-1)}{n} + o\left(\frac{t}{n}\right)\right]^n = e^{\lambda t(z-1)},$$

which is just the generating function of a Poisson distribution with parameter $\alpha = \lambda t$. Thus, X is distributed as

$$\text{Prob}\{X(t) = k\} = \frac{(\lambda t)^k}{k!} e^{-\lambda t}, \quad k = 0, 1, \ldots.$$

Observe that since $\lambda t = E[X(t)]$, the parameter λ denotes the average number of events occurring per time unit.

Exercise 6.9.3 Prove that the sum of n independent Poisson random variables X_i, $i = 1, 2, \ldots, n$, with parameters $\lambda_1, \lambda_2, \ldots, \lambda_n$, respectively, is also Poisson distributed.

Answer 6.9.3 It was seen in the text that the probability generating function of a sum of independent random variables is equal to the product of the generating functions of the random variables, i.e., that

$$G_X(z) = G_{X_1}(z)G_{X_2}(z)\cdots G_{X_n}(z),$$

where $X = X_1 + X_2 + \cdots + X_n$ and X_i, $i = 1, 2, \ldots, n$, are independent discrete random variables. It was also shown in the text that the generating function of a Poisson random variable X_i is given by

$$G_{X_i}(z) = e^{\lambda_i(z-1)}.$$

It now follows that

$$G_X(z) = e^{\lambda_1(z-1)}e^{\lambda_2(z-1)}\cdots e^{\lambda_n(z-1)} = e^{(\lambda_1+\lambda_2+\cdots+\lambda_n)(z-1)},$$

which is the generating function of a Poisson distribution with parameter $\lambda_1 + \lambda_2 + \cdots + \lambda_n$.

Exercise 6.9.4 The number of telephone calls to a central office is a Poisson process with rate $\mu = 4$ per five minute period. Find the probability that the waiting time for three or more calls to arrive is greater than two minutes. Give also the mean and standard deviation of the time to observe three calls arrive.

Answer 6.9.4 Four calls per five minutes corresponds to a rate of $\mu = 0.8$ calls per minute. Also, the probability that the waiting time for three calls or more exceeds two minutes is equal to the probability of getting 0,1 or 2 calls in two minutes. Hence

$$\begin{aligned}\text{Prob}\{W_3 > 2\} &= \text{Prob}\{N(2) \leq 2\} = e^{-1.6}\sum_{k=0}^{2}\frac{1.6^k}{k!} \\ &= 0.2019\left(1 + 1.6 + \frac{1.6^2}{2}\right) = 0.7834.\end{aligned}$$

The expectation and standard deviation are

$$E[W_3] = \frac{3}{0.8} = 3.75 \quad \text{and} \quad \sigma_{W_3} = \frac{\sqrt{3}}{0.8} = 2.17 \ \text{ minutes.}$$

Chapter 7

Continuous Probability Distributions

Exercise 7.1.1 Let X be uniformly distributed on the interval $(0, 4)$. What is the probability that the roots of $z^2 + 2Xz - 2X + 15 = 0$ are real?

Answer 7.1.1 Writing $z^2 + 2Xz - 2X + 15 = 0$ in standard form, $az^2 + bz + c = 0$, we have $a = 1$, $b = 2X$ and $c = 15 - 2X$. The roots are real if $b^2 - 4ac \geq 0$, i.e., if

$$4X^2 - 4(15 - 2X) \geq 0, \quad \text{i.e., if} \quad X^2 + 2X - 15 \geq 0,$$

which factors to

$$(X + 5)(X - 3) \geq 0.$$

Since X is uniformly distributed between 0 and 4, $X + 5$ is always positive, while $X - 3$ is non-negative if $X \geq 3$. Therefore

$$\text{Prob}\{X \geq 3\} = 1 - \text{Prob}\{X < 3\} = 0.25$$

and this is the probability that the roots are real.

Exercise 7.1.2 Let X_1 and X_2 be two independent random variables with probability density functions given by $p_{X_1}(x_1)$ and $p_{X_2}(x_2)$ respectively. Let Y be the random variable $Y = X_1 + X_2$. Find the probability density function of Y when both X_1 and X_2 are uniformly distributed on $[0, 1]$. Illustrate the probability density function of Y by means of a figure.

Answer 7.1.2 Since X_1 and X_2 are independent, their joint density function is given by

$$p_{X_1}(x_1)p_{X_2}(x_2)$$

and therefore

$$Prob\{a \leq y \leq b\} = \int\int_{a \leq x_1 + x_2 \leq b} p_{X_1}(x_1)p_{X_2}(x_2)dx_1dx_2$$

$$= \int_a^b \left(\int_{-\infty}^{\infty} p_{X_1}(y-x) p_{X_2}(x) dx \right) dy.$$

Differentiating, we find

$$p_Y(y) = \int_{-\infty}^{\infty} p_{X_1}(y-x) p_{X_2}(x) dx,$$

i.e., the convolution of p_{X_1} and p_{X_2}.

When X_1 and X_2 are uniformly distributed on $[0, 1]$, both have density functions given by

$$p(x) = \begin{cases} 1, & 0 \leq x \leq 1, \\ 0 & \text{otherwise.} \end{cases}$$

This means that the density function of Y is given by

$$p_Y(y) = \begin{cases} \int_0^y dx = y, & 0 \leq y \leq 1, \\ \int_{y-1}^1 dx = 2 - y, & 1 \leq y \leq 2, \\ 0, & y < 0 \text{ or } y > 2. \end{cases}$$

The graph of this density is triangular and is given in Figure 7.1.

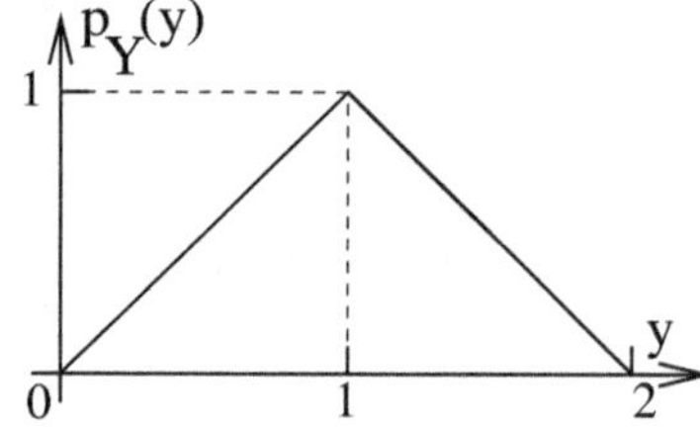

Figure 7.1: Probability density function of the random variable Y.

Exercise 7.2.1 Derive the expectation and variance of an exponential random variable from its Laplace transform.

Answer 7.2.1 The expectation and second moment are obtained by evaluating the first and second derivatives of the transform at the point zero. This gives

$$\begin{aligned} E[X] &= -\frac{dF^*(s)}{ds}\bigg|_{s=0} = \frac{\lambda}{(s+\lambda)^2}\bigg|_{s=0} = \frac{1}{\lambda} \\ E[X^2] &= (-1)^2 \frac{d^2F^*(s)}{ds^2}\bigg|_{s=0} = \frac{d}{ds}\left(\frac{-\lambda}{(s+\lambda)^2}\right)\bigg|_{s=0} = \frac{2\lambda}{(s+\lambda)^3}\bigg|_{s=0} = \frac{2}{\lambda^2}. \end{aligned}$$

The variance may now be computed as

$$\text{Var}[X] = E[X^2] - E[X]^2 = \frac{2}{\lambda^2} - \frac{1}{\lambda^2} = \frac{1}{\lambda^2}.$$

Exercise 7.2.2 The density function of an exponential random variable is given by

$$f_X(x) = \begin{cases} .5e^{-x/2}, & x > 0, \\ 0 & \text{otherwise.} \end{cases}$$

What is the expectation and standard deviation of X? Write down its cumulative distribution function and find the probability that X is greater than 4.

Answer 7.2.2 This particular exponential random variable has parameter $\lambda = 1/2$. Its expectation and standard deviation are both equal to $1/\lambda = 2$. Its cumulative distribution function is given by

$$F_X(x) = \begin{cases} 0, & x \leq 0, \\ 1 - e^{-x/2}, & x > 0. \end{cases}$$

Hence

$$\text{Prob}\{X > 4\} = 1 - \text{Prob}\{X \leq 4\} = 1 - (1 - e^{-2}) = e^{-2} = 0.1353.$$

Exercise 7.2.3 When Charles enters his local bank, he finds that the two tellers are busy, one serving Alice and the other serving Bob. Charles' service will begin as soon as the first of the two tellers becomes free. Find the probability that Charles will not be the last of Alice, Bob, and himself to leave the bank, assuming that the time taken to serve any customer is exponentially distributed with mean $1/\lambda$.

Answer 7.2.3 Since it is immaterial whether Alice or Bob finishes first, let us assume that Alice finishes, leaves and Charles begins service. Since the service time distribution has the memoryless property, the fact that Bob has already spend time being served is irrelevant. The time to serve Bob and Charles has the same distribution, so the probability that Charles will finish first is just equal to $1/2$.

Exercise 7.2.4 Let X_1 and X_2 be two independent and identically distributed exponential random variables with parameter λ. Find the density function of $Y = X_1 + X_2$ using the convolution approach.

Answer 7.2.4 The density function of both X_1 and X_2 is given by

$$f_X(x) = \begin{cases} \lambda e^{-\lambda x}, & x > 0, \\ 0 & \text{otherwise.} \end{cases}$$

The density of Y is found by means of the convolution approach. We obtain

$$f_Y(y) = \int_0^y \lambda e^{-\lambda x} \lambda e^{-\lambda(y-x)} dx = \lambda^2 e^{-\lambda y} \int_0^y e^{-\lambda x} e^{\lambda x} dx = \lambda^2 y e^{-\lambda y}.$$

Exercise 7.2.5 Show that a geometric sum of exponentially distributed random variables is itself exponentially distributed.

Hint: Let N be a geometrically distributed random variable, let $X_1, X_2, \ldots$ be independent and identically distributed exponential random variables with parameter μ, and let $S = \sum_{n=1}^{N} X_n$. Now evaluate the Laplace transform of S, $E[e^{-sS}]$, using a conditioning argument.

Answer 7.2.5 Let $X_1, X_2, \ldots$ be independent and identically distributed exponential random variables with parameter μ. Let the random variable N be geometrically distributed with probability mass function

$$p_N(n) = q(1-q)^{n-1}, \quad n = 1, 2, \ldots, \quad \text{and 0 otherwise,}$$

where $0 < q < 1$. We seek to show that $S = \sum_{n=1}^{N} X_n$ is exponentially distributed. We have

$$\begin{aligned} E[e^{-sS}] &= \sum_{k=1}^{\infty} E[e^{-sS} \,|\, N = k] \text{ Prob}\{N = k\} \\ &= \sum_{k=1}^{\infty} \left(\frac{\mu}{\mu + s} \right)^k q(1-q)^{k-1} \\ &= \frac{\mu}{\mu + s} q \sum_{k=1}^{\infty} \left(\frac{\mu(1-q)}{\mu + s} \right)^{k-1} = \frac{\mu q}{\mu + s} \sum_{k=0}^{\infty} \left(\frac{\mu(1-q)}{\mu + s} \right)^k \\ &= \frac{\mu q}{\mu + s} \frac{1}{1 - \mu(1-q)/(\mu + s)} = \frac{q\mu}{(\mu + s) - \mu(1-q)} \\ &= \frac{q\mu}{q\mu + s}, \end{aligned}$$

which is the Laplace transform for an exponentially distributed random variable with parameter μq.

Exercise 7.3.1 Show, by taking derivatives of the moment generating function of the standard normal distribution, that this distribution has mean value zero and variance equal to 1.

Answer 7.3.1 Taking derivatives of $\mathcal{M}_Z = e^{\theta^2/2}$, we obtain

$$\frac{d}{d\theta}\mathcal{M}_Z(\theta) = \theta e^{\theta^2/2}$$

and

$$\frac{d^2}{d\theta^2}\mathcal{M}_Z(\theta) = (1+\theta^2)e^{\theta^2/2}.$$

now evaluating these at $\theta = 0$ we obtain the mean and variance as

$$\mu = \left.\frac{d}{d\theta}\mathcal{M}_Z(\theta)\right|_{\theta=0} = 0$$

and

$$\sigma^2 = \left.\frac{d^2}{d\theta^2}\mathcal{M}_Z(\theta)\right|_{\theta=0} - \left(\left.\frac{d}{d\theta}\mathcal{M}_Z(\theta)\right|_{\theta=0}\right)^2 = 1.$$

Exercise 7.3.2 The time spent waiting for a bus is normally distributed with mean equal to 10 minutes and standard deviation equal to 10 minutes. Find (a) the probability of waiting less than 12 minutes and (b) the probability of waiting more than 15 minutes.

Answer 7.3.2

(a)

$$\begin{aligned}\text{Prob}\{X \le 12\} = \text{Prob}\left\{\frac{X-10}{10} \le \frac{12-10}{10}\right\} &= \text{Prob}\{Z \le 0.2\} \\ &= 0.5793.\end{aligned}$$

(b)

$$\begin{aligned}\text{Prob}\{X > 15\} &= 1 - \text{Prob}\{X \le 15\} \\ &= 1 - \text{Prob}\left\{\frac{X-10}{10} \le \frac{15-10}{10}\right\} \\ &= 1 - \text{Prob}\{Z \le 0.5\} = 1 - 0.6915 = 0.3085.\end{aligned}$$

Exercise 7.3.3 The most recent census population estimates that there are 12.5 million octogenarians in a total population of 250 million. Assume that the age (in years) of the population is normally distributed with mean equal to 38 years. Find the standard deviation of this distribution and estimate the total number of individuals older than 55.

Answer 7.3.3 Let X be the age of a random individual. We are told that

$$\text{Prob}\{X \geq 80\} = \frac{12.5}{250} = 0.05$$

and hence

$$\Phi\left(\frac{38-80}{\sigma}\right) = 1 - \Phi\left(\frac{42}{\sigma}\right) = 0.05 \quad \text{or} \quad \Phi\left(\frac{42}{\sigma}\right) = 0.95.$$

Therefore, $42/\sigma = 1.65$, i.e., $\sigma = 25.45$. We may now answer the second part of the question.

$$\begin{aligned} \text{Prob}\{X > 55\} &= 1 - \text{Prob}\{X \leq 55\} = 1 - \text{Prob}\left\{\frac{X-38}{25.45} \leq \frac{55-38}{25.45}\right\} \\ &= 1 - \text{Prob}\{Z \leq 0.6680\} = 1 - 0.7486 = 0.2514. \end{aligned}$$

The number of individuals older than 55 is then given by
$0.2514 \times 250,000,000 = 62.85$ million.

Exercise 7.4.1 Prove that the variance of a random variable having a gamma distribution with parameters α and β is given by $\text{Var}[X] = \alpha\beta^2$

Answer 7.4.1 Following the analysis in the text, we have

$$\begin{aligned} E[X^2] &= \int_0^\infty x^2 \frac{1}{\beta^\alpha \Gamma(\alpha)} x^{\alpha-1} e^{-x/\beta}\, dx = \frac{1}{\beta^\alpha \Gamma(\alpha)} \int_0^\infty x^{(2+\alpha)-1} e^{-x/\beta}\, dx \\ &= \frac{\beta^{2+\alpha}\Gamma(2+\alpha)}{\beta^\alpha \Gamma(\alpha)} \int_0^\infty \frac{1}{\beta^{2+\alpha}\Gamma(2+\alpha)} x^{(2+\alpha)-1} e^{-x/\beta}\, dx \\ &= \frac{\beta^{2+\alpha}\Gamma(2+\alpha)}{\beta^\alpha \Gamma(\alpha)} = \beta^2 \frac{(1+\alpha)\Gamma(1+\alpha)}{\Gamma(\alpha)} = \beta^2 \frac{(1+\alpha)\alpha\Gamma(\alpha)}{\Gamma(\alpha)} \\ &= \alpha(\alpha+1)\beta^2, \end{aligned}$$

and hence

$$\text{Var}[X] = E[X^2] - E[X]^2 = \alpha(\alpha+1)\beta^2 - \alpha^2\beta^2 = \alpha\beta^2.$$

Exercise 7.4.2 A random variable X has a gamma distribution with parameters $\beta = 1$ and $\alpha = 6$. Find (a) $\text{Prob}\{X \leq 5\}$, (b) $\text{Prob}\{X > 4\}$, and (c) $\text{Prob}\{4 \leq X \leq 8\}$.

Answer 7.4.2 Since $\alpha = 6$ is a positive integer, we may use the associated Poisson distribution with parameter, $x/\beta = x$.

(a)

$$\text{Prob}\{X \le 5\} = \int_0^5 \frac{1}{1^6\,\Gamma(6)} x^{6-1}e^{-x}\,dx = F(5;1,6) = 1 - \sum_{i=0}^{5}\frac{5^i}{i!}e^{-5}$$
$$= 1 - 0.6160 = 0.3840.$$

(b)

$$\text{Prob}\{X > 4\} = 1 - \int_0^4 \frac{1}{1^6\,\Gamma(6)} x^{6-1}e^{-x}\,dx = 1 - F(4;1,6)$$
$$= \sum_{i=0}^{5}\frac{4^i}{i!}e^{-4} = 0.7851.$$

(c) Since $\text{Prob}\{4 \le X \le 8\} = \text{Prob}\{X \le 8\} - \text{Prob}\{X \le 4\}$ and we already know $\text{Prob}\{X < 4\}$, we only need compute $\text{Prob}\{X \le 8\}$.

$$\text{Prob}\{X \le 8\} = \int_0^8 \frac{1}{1^6\,\Gamma(6)} x^{6-1}e^{-x}\,dx = F(8;1,6) = 1 - \sum_{i=0}^{5}\frac{8^i}{i!}e^{-8}$$
$$= 1 - 0.1912 = 0.8088.$$

Hence $\text{Prob}\{4 \le X \le 8\} = 0.8088 - (1 - 0.7851) = 0.5939$.

Exercise 7.4.3 The time it takes to eat breakfast has a gamma distribution with mean five minutes and standard deviation two minutes. Find the parameters α and β of the distribution and the probability that it takes more than eight minutes to eat breakfast.

Answer 7.4.3 We first compute the parameters, α and β, of the gamma distribution as

$$\alpha = \frac{E[X]^2}{\text{Var}[X]} = \frac{5 \times 5}{4} = 6.25 \quad \text{and} \quad \beta = \frac{\text{Var}[X]}{E[X]} = \frac{4}{5} = 0.8.$$

Since α is not a positive integer, we need to use tables of the incomplete gamma function. We find

$$\text{Prob}\{X > 8\} = 1 - \text{Prob}\{X \le 8\} = 1 - \text{Prob}\{Y \le 8/0.8\} = 1 - \text{Prob}\{Y \le 10\}.$$

From tables of the incomplete Gamma function, we find that

$$\begin{aligned} \text{Prob}\{Y \le 10\} &= 0.933, \quad \text{for } \alpha = 6, \\ \text{Prob}\{Y \le 10\} &= 0.870, \quad \text{for } \alpha = 7, \end{aligned}$$

for which we may estimate that

$$\text{Prob}\{Y \le 10\} \approx 0.917, \quad \text{for } \alpha = 6.25,$$

which gives $\text{Prob}\{X > 8\} \approx 0.083$.

Exercise 7.6.1 Derive the second moment of a Weibull random variable and from it show that

$$\text{Var}[X] = \eta^2 \left[\Gamma\left(1 + \frac{2}{\beta}\right) - \Gamma\left(1 + \frac{1}{\beta}\right)^2 \right].$$

Answer 7.6.1 Following the procedure developed in the text, we have

$$E[X^2] = \int_0^\infty x^2 \frac{\beta}{\eta^\beta} x^{\beta-1} e^{-(x/\eta)^\beta}\, dx = \frac{\beta}{\eta^\beta} \int_0^\infty e^{-(x/\eta)^\beta} x^{\beta+1}\, dx.$$

Making the substitution $u = (x/\eta)^\beta$, (or $x = \eta u^{1/\beta}$), we find $x^{\beta-1}dx = (\eta^\beta/\beta)du$ and so

$$x^{\beta+1}dx = x^2 x^{\beta-1}dx = \eta^2 u^{2/\beta} \frac{\eta^\beta}{\beta} du = \frac{\eta^{\beta+2}}{\beta} u^{2/\beta}\, du.$$

Hence

$$E[X^2] = \frac{\beta}{\eta^\beta} \int_0^\infty e^{-u} \frac{\eta^{\beta+2}}{\beta} u^{2/\beta} du = \eta^2 \int_0^\infty u^{2/\beta} e^{-u} du = \eta^2\, \Gamma\left(1 + \frac{2}{\beta}\right).$$

Finally,

$$\begin{aligned} \text{Var}[X] &= E[X^2] - E[X]^2 = \eta^2\, \Gamma(1 + 2/\beta) - (\eta\, \Gamma(1 + 1/\beta))^2 \\ &= \eta^2 \left[\Gamma\left(1 + \frac{2}{\beta}\right) - \Gamma\left(1 + \frac{1}{\beta}\right)^2 \right]. \end{aligned}$$

Exercise 7.6.2 Let the lifetime of component i, $i = 1, 2, \ldots, n$, in a series of independent components have a Weibull distribution with parameters η_i, $i = 1, 2, \ldots, n$, and β. Show that the lifetime of the complete system has a Weibull distribution and find its parameters.

Answer 7.6.2 The reliability of component i is $R_i(x) = e^{-(x/\eta_i)^\beta}$ and the reliability of the overall system is

$$\begin{aligned} R_X(x) = \prod_{i=1}^n e^{-(x/\eta_i)^\beta} &= e^{-(x/\eta_1)^\beta} e^{-(x/\eta_2)^\beta} \cdots e^{-(x/\eta_n)^\beta} \\ &= e^{-(x/\eta_1)^\beta - (x/\eta_2)^\beta - \cdots - (x/\eta_n)^\beta} \\ &= e^{-x^\beta [\eta_1^{-\beta} + \eta_2^{-\beta} + \cdots + \eta_n^{-\beta}]} = e^{-x^\beta \eta^{-\beta}} = e^{-(x/\eta)^\beta}, \end{aligned}$$

where $\eta^{-\beta} = \eta_1^{-\beta} + \eta_2^{-\beta} + \cdots + \eta_n^{-\beta}$, i.e., a Weibull distribution with parameters η and β.

Exercise 7.6.3 The time to failure of each of two embedded computers in a time-critical space mission is an exponential random variable with mean time to failure equal to 5,000 hours. At least one of the computers must function for a successful mission outcome. What is the longest time that the mission can last so that the probability of success is 0.999?

Answer 7.6.3 The reliability of each computer is $R_{X_i}(t) = e^{-t/5}$ and since they operate in parallel mode, the reliability of the system is

$$R_X(t) = 1 - (1 - e^{-t/5})^2 = 2e^{-t/5} - e^{-2t/5}.$$

This is the probability that the system will survice until time t so the question calls for us to find the value of t for which $2e^{-t/5} - e^{-2t/5} = 0.999$. Setting $z = e^{-t/5}$, we obtain the quadratic equation

$$z^2 - 2z + 0.999 = 0$$

with roots

$$\frac{2 \pm \sqrt{4 - 4(0.999)}}{2} = 1 \pm 0.03162.$$

Of the two roots 1.03162 and 0.9684, we must choose the one that is less than 1: the other gives meaningless results. We now have $e^{-t/5} = 0.9684$ or $t/5 = \ln(1/0.9684) = 0.0321$, i.e., $t = 0.1606$ which translates into a duration of 160.6 hours.

Exercise 7.6.4 Consider a system that contains four components with reliability functions given respectively by

$$R_1(t) = e^{-\alpha t}, \quad R_2(t) = e^{-\beta t}, \quad R_1(t) = e^{-\gamma t}, \quad \text{and} \quad R_1(t) = e^{-\delta t}.$$

What is the reliability function of the system that has these four components arranged (a) in series and (b) in parallel? A different system arranges these components so that the first two are in series, the last two are in series, but the two groups of two are in parallel with each other. What is the reliability function in this case?

Answer 7.6.4 When arranged in series, the reliability function of the system is

$$R_{X_s}(t) = e^{-\alpha t}\, e^{-\beta t}\, e^{-\gamma t}\, e^{-\delta t} = e^{-t(\alpha+\beta+\gamma+\delta)}.$$

When arranged in parallel, it is

$$R_{X_p}(t) = 1 - (1 - e^{-\alpha t})(1 - e^{-\beta t})(1 - e^{-\gamma t})(1 - e^{-\delta t}).$$

For the mixed arrangement, the first subsystem consisting of components 1 and 2 are in series, so

$$R_{X_{12}}(t) = e^{-\alpha t}\, e^{-\beta t} = e^{-t(\alpha+\beta)}.$$

Similarly,

$$R_{X_{34}}(t) = e^{-\gamma t}\, e^{-\delta t} = e^{-t(\gamma+\delta)}.$$

Since these two subsystems are arranged in parallel, we have the overall reliability function

$$R_X(t) = 1 - \left(1 - e^{-t(\alpha+\beta)}\right)\left(1 - e^{-t(\gamma+\delta)}\right).$$

Exercise 7.6.5 The hazard rate of a component is given by

$$h_X(t) = \frac{\alpha}{2}t^{-1/2} + \frac{\beta}{4}t^{-3/4}.$$

Find the reliability function of the component.

Answer 7.6.5 We have

$$\begin{aligned} H_X(t) &= \int_0^t \left(\frac{\alpha}{2}u^{-1/2} + \frac{\beta}{4}u^{-3/4}\right) du = \alpha \int_0^t \frac{1}{2}u^{-1/2}du + \beta \int_0^t \frac{1}{4}u^{-3/4}du \\ &= \alpha t^{1/2} + \beta t^{1/4}, \end{aligned}$$

and so the reliability is given by

$$R_X(t) = e^{-H_X(t)} = e^{-(\alpha t^{1/2} + \beta t^{1/4})}.$$

Exercise 7.6.1 Prove that the first and second moments of a hyperexponential-2 random variable X, with parameters μ_1, μ_2, and α_1, are respectively given by

$$E[X] = \frac{\alpha_1}{\mu_1} + \frac{\alpha_2}{\mu_2} \quad \text{and} \quad E[X^2] = \frac{2\alpha_1}{\mu_1^2} + \frac{2\alpha_2}{\mu_2^2},$$

where $\alpha_2 = 1 - \alpha_1$.

Answer 7.6.1 The Laplace transform of X is

$$\mathcal{L}_X(s) = \alpha_1 \frac{\mu_1}{s+\mu_1} + \alpha_2 \frac{\mu_2}{s+\mu_2},$$

and we know that

$$E[X^k] = (-1)^k \frac{d^k}{ds^k}\mathcal{L}_x(s)\bigg|_{s=0}.$$

Therefore

$$\begin{aligned} E[X] &= -1\frac{d}{ds}\left(\alpha_1 \frac{\mu_1}{s+\mu_1} + \alpha_2 \frac{\mu_2}{s+\mu_2}\right)\bigg|_{s=0} \\ &= \left(\frac{\alpha_1\mu_1}{(s+\mu_1)^2} + \frac{\alpha_2\mu_2}{(s+\mu_2)^2}\right)\bigg|_{s=0} = \frac{\alpha_1}{\mu_1} + \frac{\alpha_2}{\mu_2} \end{aligned}$$

and

$$
\begin{aligned}
E[X^2] &= (-1)^2 \frac{d^2}{ds^2}\left(\alpha_1 \frac{\mu_1}{s+\mu_1} + \alpha_2 \frac{\mu_2}{s+\mu_2}\right)\Bigg|_{s=0} \\
&= \frac{d}{ds}\left(-\alpha_1 \frac{\mu_1}{(s+\mu_1)^2} - \alpha_2 \frac{\mu_2}{(s+\mu_2)^2}\right)\Bigg|_{s=0} \\
&= \left(\frac{2\alpha_1\mu_1}{(s+\mu_1)^3} + \frac{2\alpha_2\mu_2}{(s+\mu_2)^3}\right)\Bigg|_{s=0} = \frac{2\alpha_1}{\mu_1^2} + \frac{2\alpha_2}{\mu_2^2}.
\end{aligned}
$$

Exercise 7.6.2 Derive the expectation of the Coxian-r random variable that is represented graphically in Figure 7.28 Show that the answer you compute using the notation of Figure 7.28 is the same as that derived with reference to Figure 7.27 of the text.

Answer 7.6.2 To find the expectation we proceed as follows. With probability p_1 only a single exponential phase with parameter $1/\mu_1$ is executed. With probability p_2, the expectation is that of a hypoexponential random variable with 2 phases and parameters μ_1 and μ_2; with probability p_3, the expectation is that of a hypoexponential random variable with 3 phases and parameters μ_1, μ_2 and μ_3, and so on. This implies that the expectation of the Cox-r random variable is

$$
E[X] =
$$

$$
p_1\frac{1}{\mu_1} + p_2\left(\frac{1}{\mu_1}+\frac{1}{\mu_2}\right) + p_3\left(\frac{1}{\mu_1}+\frac{1}{\mu_2}+\frac{1}{\mu_3}\right) + p_r\left(\frac{1}{\mu_1}+\frac{1}{\mu_2}+\cdots+\frac{1}{\mu_r}\right)
$$

$$
= \frac{1}{\mu_1}\sum_{i=1}^{r} p_i + \frac{1}{\mu_2}\sum_{i=2}^{r} p_i + \cdots + \frac{1}{\mu_2}\sum_{i=r}^{r} p_r
$$

$$
= \frac{1}{\mu_1} + (1-p_1)\frac{1}{\mu_2} + (1-p_1-p_2)\frac{1}{\mu_3} + \cdots + (1-p_1-p_2-\cdots-p_{r-1})\frac{1}{\mu_r}.
$$

Now observe that

$$
\begin{aligned}
1-p_1 &= \alpha_1, \\
1-p_1-p_2 &= \alpha_1 - p_2 = \alpha_1 - (1-\alpha_2)\alpha_1 = \alpha_1\,\alpha_2, \\
1-p_1-p_2-p_3 &= \alpha_1\,\alpha_2 - p_3 = \alpha_1\alpha_2 - (1-\alpha_3)\alpha_1\alpha_2 = \alpha_1\,\alpha_2\,\alpha_3,
\end{aligned}
$$

and an inductive argument shows that, in general, the coefficient of $1/\mu_k$, for $k>1$, is

$$
1 - \sum_{j=1}^{k-1} p_i = \prod_{j=1}^{k-1} \alpha_j \equiv A_k, \quad k > 1.
$$

Setting $A_1 = 1$ we obtain the same result given in the text, namely,

$$
E[X] = \sum_{k=1}^{r} \frac{A_k}{\mu_k}.
$$

Chapter 8

Bounds and Limit Theorems

Exercise 8.1.1 Let ten seconds be the mean time between the arrival of electronic orders to purchase stocks. The orders are (almost) instantaneously routed by a clerk from his computer to a trader on the stock floor. The clerk would like to go next door to buy a sandwich for lunch, but expects that it will take about two minutes. What can be said about the probability of no orders arriving during the time the clerk is off buying lunch?

Answer 8.1.1 Inserting $E(X) = 10$ and t $= 120$ into the Markov inequality, we find

$$\text{Prob}\{X \geq 120\} \leq \frac{10}{120} = 0.0833.$$

which is an upper bound on the probability that no orders arrive during the time that the clerk is gone.

Exercise 8.1.2 Construct an example for which the upper bound obtained by the Markov inequality gives a value in excess of 1.0 (and therefore serves no useful purpose, since all probabilities must be less than or equal to 1).

Answer 8.1.2 The Markov inequality will give an upper bound on the probability that is greater than one in any scenario for which the required value of t is less than the mean value. For example, using the text example of children in Centennial Campus Middle School, an upper bound on the probability of a randomly selected student being 13 years or less give an upper bound equal to $13.5/13 > 1$.

Exercise 8.1.3 Example 8.3 of the text gave the moment generating function of a normally distributed random variable with mean value $\mu = 4$ and variance $\sigma^2 = 1$ as

$$\mathcal{M}_X = e^{\mu\theta+\sigma^2\theta^2/2} = e^{4\theta+\theta^2/2},$$

and showed that the Chernoff bound yields

$$\text{Prob}\{X \geq 8\} \leq \min_{\theta \geq 0} e^{-8\theta} e^{4\theta + \theta^2/2} = \min_{\theta \geq 0} e^{(\theta^2 - 8\theta)/2}.$$

Complete this example by finding the bounds obtained from the Markov and Chebychev inequalities as well as the exact value of the probability.

Answer 8.1.3 In this example, the random variable X is normally distributed with mean $\mu = 4$ and variance $\sigma^2 = 1$. Introducing these values into the Markov and Chebychev inequalities, we respectively obtain

$$\text{Prob}\{X \geq 8\} \leq \frac{E[X]}{8} = 1/2$$

and

$$\text{Prob}\{X \geq 8\} = \text{Prob}\{|X - 4| \geq 4\} \leq \frac{\sigma_X^2}{4^2} = 1/16.$$

The exact value of the probability is given by

$$\text{Prob}\{X \geq 8\} = 1 - \text{Prob}\{X < 8\} = 1 - \text{Prob}\{Z < 4\}$$

$$= 1 - 0.9999683 = 0.0000317.$$

Exercise 8.1.4 The probability of an event occurring in one trial is 0.5. Use Chebychev's inequality to show that the probability of this event occurring between 450 and 550 times in 1000 independent trials exceeds 0.90.

Answer 8.1.4 Observe that a sequence of 1000 Bernoulli trials has a binomial distribution with mean $np = 1000/2 = 500$ and variance $np(1-p) = 250$. From Chebychev's inequality we obtain

$$\text{Prob}\{|X - 500| > 50\} \leq \frac{250}{50^2} = 0.10.$$

Exercise 8.1.5 The local garden shop keeps an average of 60 rose bushes in stock with variance equal to 64. A customer arrives and wishes to purchase 36 rose bushes. Use Chebychev's inequality to determine the likelihood of this request being met.

Answer 8.1.5 Since the variance is 64, the standard deviation is 8. Thus the question becomes that of determining the probability that the number in stock is within 3 standard deviations ($24 = 60 - 36$) of the mean. Let X be the number of rose bushes in stock. From Chebychev's inequality

$$\text{Prob}\{|X - E[X]| \leq c\sigma_X\} \geq 1 - \frac{1}{c^2},$$

we obtain

$$\text{Prob}\{|X - 60| \le 3 \times 8\} \ge 1 - \frac{1}{9} = 0.8889.$$

This is the probability that X is within 3 standard deviations of its mean value, and the probability that the customer can be satisfied. In fact, it is the probability that *any* distribution and no just the one we given above, is within 3 standard deviations of its mean.

Exercise 8.1.6 Let X be an exponentially distributed random variable with parameter $\lambda = 1$. Compare the upper bound on the probability $\text{Prob}\{X > 4\}$ obtained from the Chebychev inequality and the exact value of this probability.

Answer 8.1.6 The mean and variance of this random variable are given respectively by

$$E[X] = 1/\lambda = 1 \quad \text{and} \quad \text{Var}[X] = 1/\lambda^2 = 1.$$

From the Chebychev inequality, we obtain

$$\text{Prob}\{X > 4\} = \text{Prob}\{|X - 1| > 3\} \le 1/3^2 = 0.1111.$$

The exact value is obtained as

$$\text{Prob}\{X > 4\} = 1 - \text{Prob}\{X \le 4\} = 1 - F_X(4) = 1 - (1 - e^{-4}) = e^{-4} = 0.0183.$$

Exercise 8.2.1 Show that the weak law of large numbers provides justification for the frequency interpretation of probability.

Answer 8.2.1 Let a probability experiment be conducted a large number of times, say n, and let the random variable X_i, $i = 1, 2, \ldots, n$ have the value 1 if event $\mathcal{A}$ occurs on the ith trail and have the value 0 otherwise. The random variable $S_n = X_1 + X_2 + \cdots + X_n$, the sum of these n random variables represents the total number of times that event $\mathcal{A}$ occurred during these n trials and S_n/n is the fraction of times that $\mathcal{A}$ occurred. From the weak law of large numbers, we have

$$\lim_{n\to\infty} \text{Prob}\left\{\left|\frac{S_n}{n} - E[X]\right| \ge \epsilon\right\} = 0,$$

and hence, with high probability, S_n/n is close to $E[X]$ which in turn is equal to $\text{Prob}\{\mathcal{A}\}$.

Exercise 8.3.1 Apply the central limit theorem to compute an approximation to the probability sought in Problem 8.1.5 of the text. How does this approximation compare to the exact answer and the Chernoff bound? Recall that in Problem 8.1.5, X_i, $i = 1, 2, 3, 4, 5$, are independent exponentially distributed random variables each with the same parameter $\mu = 1/4$ and $X = \sum_{i=1}^{5} X_i$; the probability sought is $\text{Prob}\{X > 40\}$.

Answer 8.3.1 To apply the central limit theorem, we need the expectation and variance of $X = \sum_{i=1}^{5} X_i$ where the random variables X_i are independent and identically exponentially distributed with mean $1/\mu = 4$. We have

$$E[X] = 5 \times 4 = 20 \quad \text{and} \quad \text{Var}[X] = 5 \times 16 = 80.$$

Then, from the central limit theorem, we obtain

$$\begin{aligned} \text{Prob}\{X > 40\} &= \text{Prob}\left\{\frac{X-20}{\sqrt{80}} > \frac{40-20}{8.9443}\right\} \\ &= 1 - \text{Prob}\left\{\frac{X-20}{\sqrt{80}} \le 2.2361\right\} = 1 - 0.9875 = 0.0125. \end{aligned}$$

Thus the central limit approximation is closer to the true value (0.02925) than the Chernoff bound (0.2156). The interested reader may wish to examine the case when only two or three exponential random variables are added together instead of the five in this example. The approximation obtained by the central limit theorem is less satisfactory in those instances.

Exercise 8.3.2 Consider a system in which parts are routed by conveyor belts from one workstation to the next. Assume the time spent at each workstation is uniformly distributed on the interval $[0, 4]$ minutes and that the time to place an object on the conveyor belt, transport it to and then remove it from the next station, is a constant time of 15 seconds. An object begins the process by being placed on the first conveyor belt which directs it to the first station and terminates immediately after being served at the eighth workstation. Find

- the mean and variance of the time an object spends in the system,
- the probability that the total time is greater than 25 minutes, and
- the probability that the total time is less than 20 minutes.

Answer 8.3.2 Let X be the random variable that denotes the time spent on a conveyor belt plus the time at the workstation to which the object is conveyed. Then

- $E[X] = 0.25 + 2.0 = 2.25$ minutes and $\text{Var}[X] = (4-0)^2/12 = 4/3$ (since the variance of the time spent on the conveyor belt is zero). Let $X_n = 8X$ be the random variable of the total time spent in the system. It follows that $E[X_n] = 8 \times 2.25 = 18$ and $\text{Var}[X_n] = 8 \times 4/3 = 32/3$.
- $\text{Prob}\{X_n > 25\} = 1 - \Phi\left(\frac{25-18}{\sqrt{32/3}}\right) = 1 - \Phi(2.1433) = 1 - 0.9838 = 0.0162.$
- $\text{Prob}\{X_n < 20\} = \Phi\left(\frac{20-18}{\sqrt{32/3}}\right) = \Phi(0.6124) = 0.7291.$

Chapter 9

Markov Chains

Exercise 9.1.1 A machine can be in one of two states, working (denoted by s_0), or undergoing repair (denoted by s_1). Each day if it is working, it may break down with some probability that is independent of other days. It takes r days to repair, where $r > 1$. Let X_n be the state of the machine on day n.
(a) Show that X_n is not a Markov chain.
(b) Now let the state space be $S = \{s_0, s_1, \ldots, s_r\}$ where $X_n = s_i$ if the machine has been in repair for i days. Prove that X_n in this case is a Markov chain.

Answer 9.1.1 (a) Notice that

$$\text{Prob}\{X_{n+1} = s_0 \,|\, X_n = s_1,\ X_{n-1} = s_0\} = 0$$

and

$$\text{Prob}\{X_{n+1} = s_0 \,|\, X_n = \ X_{n-1} = \cdots\ X_{n-r+1} = s_1\} = 1$$

which implies that the memoryless property is not satisfied.
(b)

$$\text{Prob}\{X_{n+1} = s_{i+1} \,|\, X_n = s_i,\ \ldots\} = 1, \quad \text{if } \ 1 \leq i \leq r-1$$

and

$$\text{Prob}\{X_{n+1} = s_0 \,|\, X_n = s_r,\ \ldots\} = 1.$$

Now, no matter which sequence of states was traversed, where we are at step $n+1$ depends only on the state occupied at step n.

Exercise 9.1.2 A father buys a small goldfish for his young son who is put in charge of feeding it. Unfortunately, sometimes the son forgets to feed the fish, and then sometimes he feeds it too much with the result that when the father returns form work he finds, with probability p, independent of all previous days, that the fish is dead. In these cases, he immediately rushes to the pet store, buys an identical goldfish and replaces the dead one with the new one so that his son never finds out that his fish has died. On the father's return from work on day n, let X_n be the number of days since he last had to replace the fish. Show that X_n is a Markov chain. Find the single step transition

probability p_{ik} and the n-step transition probability $p_{ik}^{(n)}$.
Notice that this Markov chain has an infinite state space.

Answer 9.1.2 Notice that

$$X_{n+1} = \begin{cases} X_n + 1 & \text{with probability } 1-p \\ 0 & \text{with probability } p. \end{cases}$$

Since p is independent of previous days, it must follow that X_n is a Markov chain. The single step transition probabilities are

$$p_{ik} = \begin{cases} 1-p, & \text{if } k = i+1 \\ p, & \text{if } k = 0 \\ 0 & \text{otherwise} \end{cases}$$

To find the n-step transition probabilities, we note that the goldfish either survived for all n days, or else it has died sometime during that period. Hence

$$p_{ik}^{(n)} = (1-p)^n \quad \text{if } k = i+n$$

and

$$p_{ik}^{(n)} = (1-p)^k p \quad \text{if } 0 \le k \le n-1.$$

The fish dies once, with probability p from which time we still have to take k steps (each with probability $1-p$) to finish off.

Exercise 9.1.3 A stack of N books lies on a desk. You take a book from the stack, read what you need and then place the book on the top of the stack. How would you represent this as a Markov chain? What are the states? What transitions are possible? Give the transition probability matrix when $N = 3$ and under the assumption that a book is picked from the stack at random.

Answer 9.1.3 Let each of the $N!$ permutations of the first N integers be a state, then this scenario becomes a Markov chain if each choice of a book is independent of all prior choices. Transitions are possible from a given permutation to that same permutation, if the chosen book happens to be at the top of the stack, or to a permutation in which the chosen index becomes the first in the permutation and all indices prior to the chosen index move down one position.
With $N = 3$ there are 6 permutations which we label

$$123, \quad 132, \quad 213, \quad 231, \quad 312, \quad \text{and } 321.$$

The transition matrix is then

$$P = \begin{pmatrix} 1/3 & 0 & 1/3 & 0 & 1/3 & 0 \\ 0 & 1/3 & 1/3 & 0 & 1/3 & 0 \\ 1/3 & 0 & 1/3 & 0 & 0 & 1/3 \\ 1/3 & 0 & 0 & 1/3 & 0 & 1/3 \\ 0 & 1/3 & 0 & 1/3 & 1/3 & 0 \\ 0 & 1/3 & 0 & 1/3 & 0 & 1/3 \end{pmatrix}$$

Exercise 9.1.4 Consider a store that sells television sets. If at the end of the day there is one or zero sets left, then that evening, after the store has closed, the shopkeeper brings in enough new sets so that the number of sets in stock for the next day is equal to five. This means that each morning, at store opening time, there are between two and five television sets available for sale. Such a policy is said to be an (s, S) inventory control policy. Here we have assigned the values $s = 1$, $S = 5$. The shopkeeper knows from experience that the probabilities of selling 0 through 5 sets on any given day are 0.4, 0.3, 0.15, 0.15, 0.0, and 0.0.

Explain how this scenario may be modeled by a Markov chain $\{X_n,\ n = 1, 2, \ldots\}$, where X_n is the random variable that defines the number of television sets left at the end of the nth day. Write down and explain the structure of the transition probability matrix.

Answer 9.1.4 The number left at the end of any day just depends on the number at the end of the previous day. Row number $i = 0, 1, \ldots, 5$ corresponds to i television sets left the previous day.

$$P = \begin{pmatrix} 0 & 0 & 0.15 & 0.15 & 0.3 & 0.4 \\ 0 & 0 & 0.15 & 0.15 & 0.3 & 0.4 \\ 0.3 & 0.3 & 0.4 & 0 & 0 & 0 \\ 0.15 & 0.15 & 0.3 & 0.4 & 0 & 0 \\ 0 & 0.15 & 0.15 & 0.3 & 0.4 & 0 \\ 0 & 0 & 0.15 & 0.15 & 0.3 & 0.4 \end{pmatrix}.$$

If at the end of the previous day there are only 0 or 1 sets available a sufficient number of sets are brought into the store to raise this number to 5. With probabilities 0.4, 0.3, 0.15 and 0.15, this number is decreased by 0, 1, 2 and 3 respectively at the end of the day.

If the number left behind at the end of the previous day is greater than or equal to three then the inventory is not increased and there is enough to satisfy all of the current day's demand. At the end of the current day, the number of televisions sold may be equal to 0, 1, 2 or 3 and transitions into the corresponding states occur with probabilities 0.4, 0.3, 0.15 and 0.15 respectively.

When at the start of the day, there are only two television sets in the store, the store owner has no problem in selling 0, 1, or 2 sets. However, there is a nonzero probability that he could have sold 3 sets that day, i.e., there is a possibility that one customer wishing to purchase a set goes away empty handed. In this case the probability of going from state 2 to state 0 is equal to 0.3 (the sum of the probabilities of selling 2 or 3 sets).

Exercise 9.1.5 William, the collector, enjoys collecting the toys in McDonald's Happy Meals. And now McDonald's has come out with a new collection containing five toy warriors. Each Happy Meal includes one randomly chosen warrior. Naturally William has to collect all five different types.

(a) Use a discrete-time Markov chain to represent the process that William will go through to collect all five warriors and draw the state transition diagram.

(b) Construct the stochastic transition probability matrix for this discrete-time Markov chain and compute the probability distribution after William has eaten three happy meals.

(c) Let T denote the total number of Happy Meals that William will eat to enable him to get all five warriors. Compute $E[T]$ and $\text{Var}[T]$.

Answer 9.1.5

(a) Let state i represent the situation in which William has collected i different warriors. Thus there are a total of 5 states. On buying his first happy meal, William enters state 1. Since William believes (and we assume) that the warriors are randomly distributed among the happy meals, the probability of moving into state 2 is given by 4/5, while the probability of remaining in state 1 is given by 1/5. The same reasoning allows us to draw the following transition probability diagram.

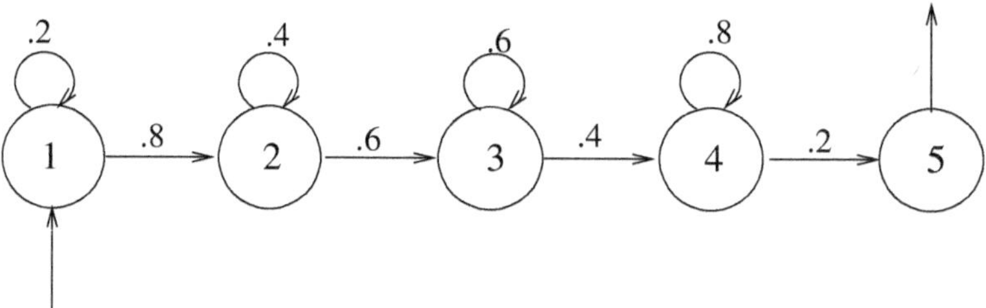

Figure 9.1: State transition diagram.

- The matrix of transition probabilities is given by

$$P = \begin{pmatrix} .2 & .8 & 0 & 0 & 0 \\ 0 & .4 & .6 & 0 & 0 \\ 0 & 0 & .6 & .4 & 0 \\ 0 & 0 & 0 & .8 & .2 \\ 0 & 0 & 0 & 0 & 1 \end{pmatrix}.$$

The initial vector is given by

$$\pi_0 = (1, 0, 0, 0, 0).$$

After eating his first meal, William gets his first warrior and is thus in state 1, the initial state. After eating 2, the probability distribution is given by $\pi_0 P = (.2, .8, 0, 0, 0)$ and after eating 3, it is given by $(.04, .48, .48, 0, 0)$.

- To answer the third part, we need to determine the number of meals that William eats in states 1 through 4. Let these define random variables T_i, $i = 1, 2, 3, 4$. Thus, for example, T_3 denotes the number of meals that William must eat before he gets his fourth warrior. In a more general context, T_i represents the sojourn time in state i. This is geometrically distributed with parameter $p_i = (5-i)/5$. (View it as a sequence of Bernoulli trials with probability of success equal to p_i.) We have

$$E[T_1] = \frac{1}{p_1} = \frac{5}{4}, \quad E[T_2] = \frac{1}{p_2} = \frac{5}{3}, \quad E[T_3] = \frac{1}{p_3} = \frac{5}{2}, \quad E[T_4] = \frac{1}{p_4} = \frac{5}{1}.$$

and hence

$$E[T] = 1 + \sum_{i=1}^{4} E[T_i] = 1 + \frac{125}{12} = 11.4167.$$

Also, since the T_i are independent, we can compute the variance as

$$\text{Var}[T] = \sum_{i=1}^{4} \text{Var}[T_i] = \sum_{i=1}^{4} \frac{1-p_i}{p_i^2} = \frac{.2}{.8^2} + \frac{.4}{.6^2} + \frac{.6}{.4^2} + \frac{.8}{.2^2} = 25.1736.$$

Exercise 9.2.1 The following matrix is the single-step transition probability matrix of a discrete time Markov chain which describes the weather. State 1 represents a sunny day, state 2 a cloudy day, and state 3 a rainy day.

$$P = \begin{pmatrix} 0.7 & 0.2 & 0.1 \\ 0.3 & 0.5 & 0.2 \\ 0.2 & 0.6 & 0.2 \end{pmatrix}$$

(a) What is the probability of a cloudy day being followed by two sunny days?

(b) Given that today the sun shines, what is the probability that it will rain the day after tomorrow?

(c) What is the mean length of a cloudy period?

Answer 9.2.1 (a) $\text{Prob}\{C \to S \to S\} = 0.3 \times 0.7 = 0.21$.

(b) Given that today is rainy, the probability that the sun will shine the day after tomorrow is given by the SR (i.e., 13) element of P^2.

$$\begin{aligned} p_{SR}^{(2)} &= 0.7 \times 0.1 + 0.2 \times 0.2 + 0.1 \times 0.2 \\ &= 0.07 + 0.04 + 0.02 \\ &= 0.13 \end{aligned}$$

(c) The mean length of a cloudy period is given by $1/(1 - p_{CC}) = 1/0.5 = 2$.

Exercise 9.2.2 Consider the four-state discrete-time Markov chain whose transition probability matrix at time step n, $n = 0, 1, \ldots$, is

$$P(n) = \begin{pmatrix} 0 & 0.6 & 0.4 & 0 \\ 0.8 & 0 & 0 & 0.2 \\ 0 & 0.5(0.5)^n & 0 & 1-0.5(0.5)^n \\ 0 & 0 & 0.8(0.8)^n & 1-0.8(0.8)^n \end{pmatrix}.$$

What is the probability distribution after three steps if the Markov chain is initiated in (a) state 1; (b) state 4?

Answer 9.2.2 The probability distributions after three steps are computed from $\pi^{(0)}P(0)P(1)P$ We first compute $P(0)P(1)P(2)$, where

$$P(0) = \begin{pmatrix} 0 & 0.6 & 0.4 & 0 \\ 0.8 & 0 & 0 & 0.2 \\ 0 & 0.5 & 0 & 0.5 \\ 0 & 0 & 0.8 & 0.2 \end{pmatrix}, \quad P(1) = \begin{pmatrix} 0 & 0.6 & 0.4 & 0 \\ 0.8 & 0 & 0 & 0.2 \\ 0 & 0.25 & 0 & 0.75 \\ 0 & 0 & 0.64 & 0.36 \end{pmatrix},$$

and

$$P(2) = \begin{pmatrix} 0 & 0.6 & 0.4 & 0 \\ 0.8 & 0 & 0 & 0.2 \\ 0 & 0.125 & 0 & 0.875 \\ 0 & 0 & 0.512 & 0.488 \end{pmatrix}$$

We find

$$P(0)P(1)P(2) = \begin{pmatrix} 0.0800 & 0.2880 & 0.4070 & 02250 \\ 0.3840 & 0.0560 & 0.0369 & 0.5231 \\ 0.0000 & 0.2800 & 0.3034 & 0.4166 \\ 0.1600 & 0.0160 & 0.3441 & 0.4799 \end{pmatrix}$$

Therefore, if we begin in state 1, the probability distribution after three steps is

$$(1,0,0,0)\begin{pmatrix} 0.0800 & 0.2880 & 0.4070 & 02250 \\ 0.3840 & 0.0560 & 0.0369 & 0.5231 \\ 0.0000 & 0.2800 & 0.3034 & 0.4166 \\ 0.1600 & 0.0160 & 0.3441 & 0.4799 \end{pmatrix}$$

$$= (0.0800, 0.2880, 0.4070, 02250)$$

while, if we begin in state 4, we find,

$$(0,0,0,1)\begin{pmatrix} 0.0800 & 0.2880 & 0.4070 & 02250 \\ 0.3840 & 0.0560 & 0.0369 & 0.5231 \\ 0.0000 & 0.2800 & 0.3034 & 0.4166 \\ 0.1600 & 0.0160 & 0.3441 & 0.4799 \end{pmatrix}$$

$$= (0.1600, 0.0160, 0.3441, 0.4799).$$

Exercise 9.3.1 Give example state transition diagrams for the following types of Markov chain states. Justify your answers. (a) Transient state, (b) positive recurrent state, (c) periodic state, (d) absorbing state, (e) closed set of states, and (f) irreducible chain.

Answer 9.3.1
(a) Transient state:

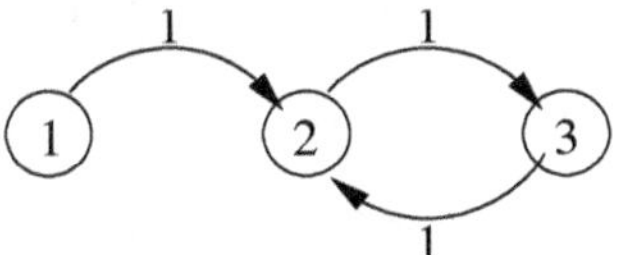

Figure 9.2: State 1 is transient.

In the Markov chain of Figure 9.2, state 1 is a transient state since

$$f_{11}^{(1)} = f_{11}^{(2)} = f_{11}^{(3)} = \cdots = 0 \quad \text{and} \quad f_{11} = \sum_{n=1}^{\infty} f_{11}^{(n)} = 0 < 1.$$

(b) Positive recurrent state: In the Markov chain of Figure 9.3, state 1 is positive recurrent since

$$f_{11}^{(1)} = f_{11}^{(2)} = 0; \quad f_{11}^{(3)} = 1; \quad f_{11}^{(4)} = f_{11}^{(5)} = f_{11}^{(6)} = \cdots = 0.$$

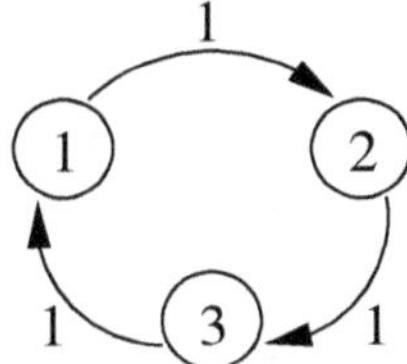

Figure 9.3: State 1 is positive recurrent.

Hence

$$M_{11} = \sum_{n=1}^{\infty} n f_{11}^{(n)} = 1 \times 0 + 2 \times 0 + 3 \times 1 + 4 \times 0 + \cdots = 3$$

is finite.
(c) Periodic state: All states in the Markov chain of Figure 9.4 are periodic with period 2. Upon leaving any state, a return is possible only in a number of steps that is a multiple of 2. Also, all states in Figure 9.3 are periodic with period 3.

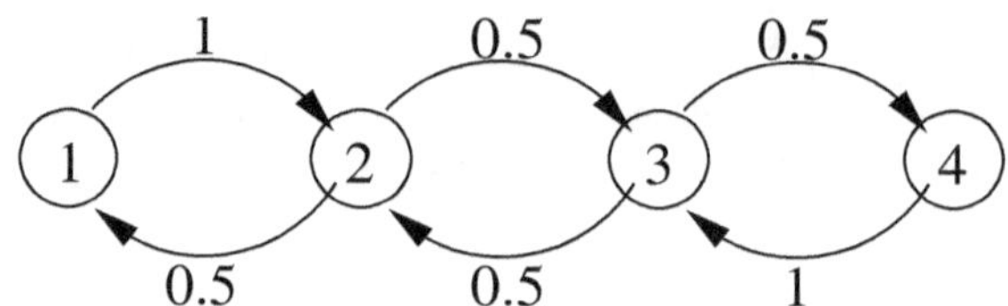

Figure 9.4: All states are periodic.

(d) and (e) Absorbing state and closed set of states: In Figure 9.5, $c_1 = \{1, 2, 3\}$; $c_2 = \{5\}$ are two closed sets. State 5 is an absorbing state, i.e., a closed set containing a single state. Once the system moves into one of these closed sets, it cannot move out of it.

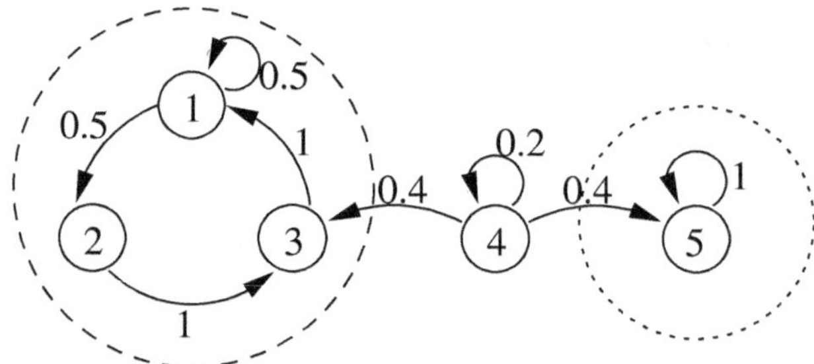

Figure 9.5: State 5 is an absorbing state.

(f) Irreducible chain: The Markov chain of Figure 9.6 is irreducible because every state is reachable from every other state. It is a single closed set, and all states are members of that set.

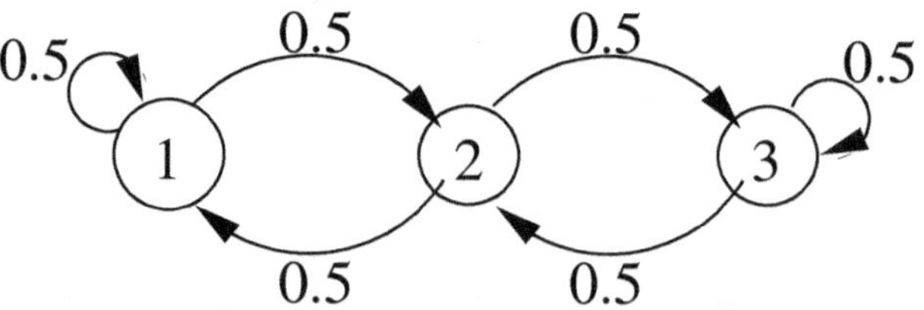

Figure 9.6: Irreducible Markov chain.

Exercise 9.3.2 Find the mean recurrence time of state 2 and the mean first passage time from state 1 to state 2 in the discrete-time Markov chain whose transition probability matrix is given by:

$$P = \begin{pmatrix} 0.7 & 0.2 & 0.1 \\ 0.3 & 0.5 & 0.2 \\ 0.2 & 0.6 & 0.2 \end{pmatrix}$$

Hint: Use Matlab.

Answer 9.3.2 Using the initial values,

$$M = \begin{pmatrix} 1 & 1 & 1 \\ 1 & 1 & 1 \\ 1 & 1 & 1 \end{pmatrix}, \quad E = \begin{pmatrix} 1 & 1 & 1 \\ 1 & 1 & 1 \\ 1 & 1 & 1 \end{pmatrix}, \quad \text{and} \quad D = Diag(M) = \begin{pmatrix} 1 & 0 & 0 \\ 0 & 1 & 0 \\ 0 & 0 & 1 \end{pmatrix}$$

and repeatedly executing (iterating) with the following scheme (in Matlab code)

$$D(1,1) = M(1,1); \quad D(2,2) = M(2,2); \quad D(3,3) = M(3,3); \quad M = E + P * (M - D)$$

we find

$$M^{(1)} = \begin{pmatrix} 1.3000 & 1.8000 & 1.9000 \\ 1.7000 & 1.5000 & 1.8000 \\ 1.8000 & 1.4000 & 1.8000 \end{pmatrix}; \quad M^{(2)} = \begin{pmatrix} 1.5200 & 2.4000 & 2.6900 \\ 2.2100 & 1.8200 & 2.4700 \\ 2.3800 & 1.6400 & 2.4600 \end{pmatrix}$$

$$M^{(3)} = \begin{pmatrix} 1.6800 & 2.8440 & 3.3770 \\ 2.5810 & 2.0480 & 3.0420 \\ 2.8020 & 1.8080 & 3.0200 \end{pmatrix}; \quad M^{(4)} = \begin{pmatrix} 1.7964 & 3.1716 & 3.9723 \\ 2.8509 & 2.2148 & 3.5341 \\ 3.1090 & 1.9304 & 3.5006 \end{pmatrix}$$

$$M^{(5)} = \begin{pmatrix} 1.8811 & 3.4132 & 4.4874 \\ 3.0473 & 2.3376 & 3.9587 \\ 3.3323 & 2.0204 & 3.9149 \end{pmatrix}; \quad M^{(6)} = \begin{pmatrix} 1.9427 & 3.5913 & 4.9329 \\ 3.1901 & 2.4280 & 4.3256 \\ 3.4948 & 2.0867 & 4.2727 \end{pmatrix}$$

$$\lim_{k \to \infty} M^{(k)} = \begin{pmatrix} 2.1071 & 4.0909 & 7.7778 \\ 3.5714 & 2.6818 & 6.6667 \\ 3.9286 & 2.2727 & 6.5556 \end{pmatrix}.$$

Therefore, the mean recurrence time of state 2 is equal to 2.6818 and the mean first passage time from state 1 to state 2 is equal to 4.0909 .

Exercise 9.4.1 In the Markov chain whose transition probability matrix P is given below, identify examples of the following (if they exist):

(a) a return state
(b) a non-return state
(c) an absorbing state
(d) a closed communicating class
(e) an open communicating class
(f) a closed communicating class containing recurrent states
(g) an open communicating class containing recurrent states
(h) a closed communicating class containing transient states
(i) an open communicating class containing transient states
(j) a communicating class with both transient and recurrent states

$$P = \begin{pmatrix} 0 & 1.0 & 0 & 0 & 0 & 0 & 0 & 0 \\ 0 & 0 & 0 & 1.0 & 0 & 0 & 0 & 0 \\ 0 & 0 & 1.0 & 0 & 0 & 0 & 0 & 0 \\ 1.0 & 0 & 0 & 0 & 0 & 0 & 0 & 0 \\ 0 & 0 & 0 & 1.0 & 0 & 0 & 0 & 0 \\ 0 & 0 & 0 & 0.5 & 0.5 & 0 & 0 & 0 \\ 0 & 0 & 0 & 0 & 0 & 0.5 & 0 & 0.5 \\ 0 & 0 & 0 & 0 & 0.5 & 0 & 0.5 & 0 \end{pmatrix}.$$

Answer 9.4.1
(a) a return state: Any state other than 5 or 6.
(b) a non-return state: 5 (or 6)
(c) an absorbing state: 3
(d) a closed communicating class: $\{1, 2, 4\}$ (or $\{3\}$)
(e) an open communicating class: $\{7, 8\}$
(f) a closed communicating class containing recurrent states: $\{1, 2, 4\}$ (or $\{3\}$)
(g) an open communicating class containing recurrent states: Does not exist
(h) a closed communicating class containing transient states: Does not exist
(i) an open communicating class containing transient states: $\{7, 8\}$
(j) a communicating class with both transient and recurrent states: Does not exist.

Exercise 9.4.2 A Markov chain with transition probability matrix P is said to be *regular* if for some $n_0 < \infty$ we have

$$p_{ij}(n_0) > 0 \text{ for all } i \text{ and } j.$$

(a) Describe in words what this means.

(b) What is the relationship between a *regular* Markov and an *irreducible* Markov chain?

(c) Give an example of a Markov chain that is irreducible but not regular, if indeed such a chain can possibly exist. If such a chain cannot exist, then explain why.

(d) Give an example of a Markov chain that is regular but not irreducible, if indeed such a chain can possibly exist. If such a chain cannot exist, then explain why.

Answer 9.4.2 (a) This means that a Markov chain is regular if there is a time such that, no matter where it started, the chain could be anywhere in S, i.e., all the elements of $P(n_0)$ are strictly positive.

(b) Recall the conditions for irreducibility:
A Markov chain is irreducible if for each i and k in S there exists an $n_0 < \infty$ such that $p_{ij}(n_0) > 0$.

Note the difference in the terminology: "*for all*" in the definition of regularity and "*for each*" in the definition of irreducibility.

(c) The (periodic) Markov chain with transition probability matrix

$$P = \begin{pmatrix} 0 & 1 \\ 1 & 0 \end{pmatrix}$$

is irreducible but it is not regular.

(d) No such example exists — irreducibility is a weaker condition than regularity. A regular Markov chain must be irreducible, but not the reverse.

Exercise 9.4.3 Consider a discrete-time Markov chain with transition probabilities given by

$$p_{ij} = e^{-\lambda} \sum_{k=0}^{j} \begin{pmatrix} i \\ k \end{pmatrix} p^k q^{i-k} \frac{\lambda^{j-k}}{(j-k)!},$$

where $p + q = 1$, $0 \le p \le 1$ and $\lambda > 0$.

(a) Is this chain reducible? Explain.

(b) Is this chain periodic? Explain.

Answer 9.4.3 Recall that

$$\begin{pmatrix} i \\ k \end{pmatrix} = \begin{cases} i!/(k!(i-k)!) & \text{for } 0 \le k \le i, \\ 0 & \text{for } k < 0 \text{ or } k > i. \end{cases}$$

(a) This Markov chain is irreducible, since $p_{ij} \neq 0$ for all $i \neq j$.
There is a path (in fact, a single-step path) from any state i to every other state j.

(b) This Markov chain is *aperiodic* since there is at least one value of i for which $p_{ii} \neq 0$.

Exercise 9.5.1 In a country at the end of each year, the government arranges a poll where each person evaluates his/her financial status as "good", "fair", or "bad". Although their individual income swill vary, we assume that for each person the process of transition from one state to another can be represented by a Markov chain with transition probability matrix given by

$$P = \begin{pmatrix} 0.6 & 0.4 & 0.0 \\ 0.2 & 0.6 & 0.2 \\ 0.0 & 0.6 & 0.4 \end{pmatrix}.$$

What percentage of citizens will transit from "fair" to "good" at statistical equilibrium?

Answer 9.5.1 Solving the global balance equations, we find the stationary probability of being in state "fair" to be $6/11 \approx 0.545$.
Since $p_{21} = 0.2$, from these 54.5% of citizens, on the average 20% will transit to the "good" state. Hence the solution is $6/11 \times 0.2\ 6/55 \approx 11\%$.

Exercise 9.5.2 A fair coin is tossed repeatedly. Let X_n be the random variable that counts the number of consecutive heads at time n since the previous tail (also called the current *run of heads*. For example, given the sequence `HHTHTTHHHT` and beginning with $X_0 = 0$ we have

$$X_1 = 1,\ X_2 = 2,\ X_3 = 0,\ X_4 = 1,\ X_5 = 0,$$
$$X_6 = 0,\ X_7 = 1,\ X_8 = 2,\ X_9 = 3,\ X_{10} = 0,$$

(1) Describe this sequence of success runs as a Markov chain and draw the state transition diagram.

(2) Find its stationary distribution

(3) Are the states of this Markov chain positive-recurrent, null-recurrent or transient?

(4) Find its limiting distribution.

Answer 9.5.2
1. The state space is $S = \{0, 1, 2, 3, \ldots\}$. The only transitions possible from state i are to state $i+1$ and to state 0, each with probability 0.5.
2. The balance equations are

$$(\pi_0, \pi_1, \ldots) \begin{pmatrix} 0.5 & 0.5 & 0 & 0 & 0 & \cdots \\ 0.5 & 0 & 0.5 & 0 & 0 & \cdots \\ 0.5 & 0 & 0 & 0.5 & 0 & \cdots \\ 0.5 & 0 & 0 & 0 & 0.5 & \cdots \\ \vdots & \vdots & \vdots & \vdots & \vdots & \ddots \end{pmatrix} = (\pi_0, \pi_1, \ldots)$$

and have the obvious solution (the geometric distribution with probability of success = 0.5):

$$\pi_i = \frac{1}{2^{i+1}}, \quad i = 0, 1, 2, \ldots$$

3. The states are all positive-recurrent since the elements of the unique stationary distribution are all strictly positive. (Notice that the Markov chain is irreducible.)
4.The Markov chain is aperiodic (note the loop on the first state) positive-recurrent and irreducible and so its stationary distribution is also its limiting distribution.

Exercise 9.5.3 Let $S = \{0, 1, 2, \ldots, N\}$ be the set of states of a Markov chain. Transitions from any state $i < N$ are to state $i + 1$ with probability p_i and to state i (a self loop) with probability $q_i = 1 - p_i$. Transitions from state N are to state 0 with probability p_N or a self loop with probability $q_N = 1 - p_N$. Find the stationary distribution of this Markov chain.

Answer 9.5.3 From the global balance equations, $\pi = \pi P$, we have

$$\begin{aligned}
\pi_0 &= q_0\pi_0 + p_N\pi_N \\
\pi_1 &= p_0\pi_0 + q_1\pi_1 \\
\pi_2 &= p_1\pi_1 + q_2\pi_2 \\
\vdots & \quad \vdots
\end{aligned}$$

which lead to

$$\begin{aligned}
p_0\pi_0 &= p_N\pi_N \\
p_1\pi_1 &= p_0\pi_0 \\
p_2\pi_2 &= p_1\pi_1 \\
\vdots & \quad \vdots
\end{aligned}$$

Solving in terms of π_0:

$$\begin{aligned}
\pi_1 &= \frac{p_0}{p_1}\pi_0 \\
\pi_2 &= \frac{p_1}{p_2}\pi_1 = \frac{p_0}{p_2}\pi_0 \\
\pi_3 &= \frac{p_2}{p_3}\pi_2 = \frac{p_0}{p_3}\pi_0 \\
\vdots & \quad \vdots \\
\pi_N &= \frac{p_0}{p_N}
\end{aligned}$$

Since we must have $\sum_{i=0}^{N} \pi_i = 1$:

$$1 = \left(1 + \frac{p_0}{p_1} + \frac{p_0}{p_2} + \cdots + \frac{p_0}{p_N}\right)\pi_0$$

i.e.,

$$p_0\pi_0 \sum_{i=0}^{N} \frac{1}{p_i} = 1$$

Let $\alpha = \sum_{i=0}^{N} 1/p_i$. It now follows that the stationary distribution is given by

$$\pi_i = \frac{1}{\alpha p_i}, \quad i = 0, 1, 2, \ldots, N.$$

Exercise 9.5.4 In the scenario of Exercise 9.2.1,

(a) What is the unconditional probability of having a sunny day?

(b) What is the mean number of rainy days in a month of 31 days?

(c) What is the mean recurrence time of sunny days?

Answer 9.5.4

(a) To find the unconditional probability of having a sunny day, let the steady-state probabilities be $(\pi_S\ \pi_C\ \pi_R)$. We know that $\pi_S + \pi_C + \pi_R = 1$, which together with

$$(\pi_S\ \pi_C\ \pi_R)\begin{pmatrix} 0.7 & 0.2 & 0.1 \\ 0.3 & 0.5 & 0.2 \\ 0.2 & 0.6 & 0.2 \end{pmatrix} = (\pi_S\ \pi_C\ \pi_R)$$

allows us to compute the (normalized) solution as

$$\left.\begin{array}{r} 0.7\pi_S + 0.3\pi_C + 0.2\pi_R = \pi_S \\ 0.2\pi_S + 0.5\pi_C + 0.6\pi_R = \pi_C \\ 0.1\pi_S + 0.2\pi_C + 0.2\pi_R = \pi_R \end{array}\right\} \Longrightarrow \left\{\begin{array}{l} \pi_S = 0.4746 \\ \pi_C = 0.3729 \\ \pi_R = 0.1525 \end{array}\right.$$

Steady-state probability of a sunny day $= \pi_S = 0.4746$.

(b) The mean number of rainy days in a month of 31 days is given by $31 \times \pi_R$, where π_R is the unconditional probability of a rainy day. It follows that the required answer is $31 \times 0.152 = 4.712$ days. (Definitely not the weather at Belfast :-)

(c) The mean recurrence time of two sunny days is obtained as $M_{SS} = 1/\pi_S = 2.019$ days.

Exercise 9.5.5 Consider a Markov chain defined on the nonnegative integers and having transition probabilities ($0 < p < 1$) given by

$$p_{n,n+1} = p \text{ and } p_{n,0} = 1 - p.$$

(a) Compute the mean recurrence time of state 0.

(b) Show that the Markov chain is positive recurrent.

(c) Prove that the limiting probabilities π exist.

(d) Find π (in terms of p).

Answer 9.5.5

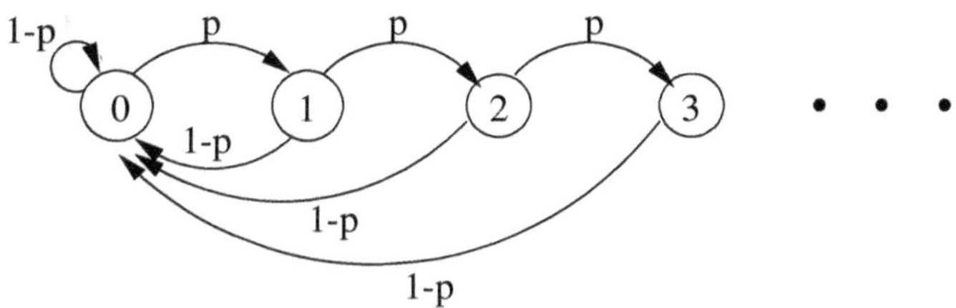

Figure 9.7: State transition diagram for Exercise 9.5.5.

(a)

$$f_{jj}^{(n)} = \text{Prob}\{\text{first return to state } j \text{ occurs } n \text{ steps after leaving it}\}$$

$$M_{jj} = \sum_{n=1}^{\infty} n f_{jj}^{(n)}.$$

$$\begin{aligned} M_{00} &= (1-p) + 2(1-p)p + 3(1-p)p^2 + \cdots \\ &= (1-p)[\,1 + 2p + 3p^2 + \cdots\,] \\ &= (1-p)\frac{d}{dp}[\,p + p^2 + p^3 + \cdots\,] \\ &= (1-p)\frac{d}{dp}\left[\frac{p}{1-p}\right] \text{(since } 0 < p < 1) \\ &= (1-p)\left[\frac{(1-p)+p}{(1-p)^2}\right] = \frac{1}{1-p}. \end{aligned}$$

(b)

$$M_{00} = \frac{1}{1-p} < \infty \Longrightarrow \text{state 0 is positive recurrent.}$$

Also, all states communicate, i.e., the matrix is irreducible. In an irreducible Markov chain, if one state is positive recurrent, then all states must be positive recurrent. Thus the Markov chain itself is positive recurrent.

(c) An irreducible positive recurrent Markov chain has a unique stationary distribution which is equal to its limiting probabilities.

(d)

$$\pi(0) = \frac{1}{M_{00}} = (1-p).$$

From $\pi = \pi P$ and since

$$P = \begin{pmatrix} 1-p & p & & & \\ 1-p & & p & & \\ 1-p & & & p & \\ 1-p & & & & p \\ \vdots & & & & \ddots \end{pmatrix},$$

$$\pi(1) = \pi(0)p = (1-p)p,$$

$$\pi(2) = \pi(1)p = (1-p)p^2,$$

$$\vdots$$

In general,

$$\pi(i) = \pi(i-1)p = (1-p)p^i.$$

Exercise 9.5.6 Let $\{X_n, n \geq 0\}$ be a two-state Markov chain, whose transition probability matrix is given by

$$P = \begin{pmatrix} 1-p & p \\ q & 1-q \end{pmatrix},$$

where $0 < p < 1$ and $0 < q < 1$. Let $\text{Prob}\{X_0 = 0\} = \pi_0(0)$ be the probability that the Markov chain begins in state 0. Prove by induction that

$$\text{Prob}\{X_n = 0\} = (1-p-q)^n \pi_0(0) + q \sum_{j=0}^{n-1} (1-p-q)^j.$$

Answer 9.5.6 Let $n = 1$;

$$\begin{aligned} \text{Prob}\{X_1 = 0\} &= \text{Prob}\{X_1 = 0 \mid X_0 = 0\}\text{Prob}\{X_0 = 0\} + \\ &\quad \text{Prob}\{X_1 = 0 \mid X_0 = 1\}\text{Prob}\{X_0 = 1\} \\ &= (1-p)\pi_0(0) + q[1 - \pi_0(0)] \\ &= (1-p-q)\pi_0(0) + q. \end{aligned}$$

Let $n = 2$;

$$\begin{aligned}
\text{Prob}\{X_2 = 0\} &= \text{Prob}\{X_2 = 0 \mid X_1 = 0\}\text{Prob}\{X_1 = 0\} + \\
&\quad \text{Prob}\{X_2 = 0 \mid X_1 = 1\}\text{Prob}\{X_1 = 1\} \\
&= (1-p)[\,(1-p-q)\pi_0(0) + q\,] \\
&\quad +q[\,1 - (1-p-q)\pi_0(0) - q\,] \\
&= (1-p-q)[\,(1-p-q)\pi_0(0) + q\,] + q \\
&= (1-p-q)^2\pi_0(0) + (1-p-q)q + q.
\end{aligned}$$

Thus the formula holds for the basis clauses. Now let us assume that

$$\text{Prob}\{X_n = 0\} = (1-p-q)^n\pi_0(0) + q\sum_{j=0}^{n-1}(1-p-q)^j$$

and prove it true for $n+1$. We shall use the relationship

$$\begin{aligned}
\text{Prob}\{X_{n+1} = 0\} &= \text{Prob}\{X_{n+1} = 0 \mid X_n = 0\}\text{Prob}\{X_n = 0\} + \\
&\quad \text{Prob}\{X_{n+1} = 0 \mid X_n = 1\}\text{Prob}\{X_n = 1\}.
\end{aligned}$$

This gives

$$\text{Prob}\{X_{n+1} = 0\} = (1-p)\text{Prob}\{X_n = 0\} + q(1 - \text{Prob}\{X_n = 0\})$$

$$\begin{aligned}
&= (1-p)\left[(1-p-q)^n\pi_0(0) + q\sum_{j=0}^{n-1}(1-p-q)^j\right] \\
&\quad + \left[q\left(1 - (1-p-q)^n\pi_0(0) - q\sum_{j=0}^{n-1}(1-p-q)^j\right)\right] \\
&= (1-p)(1-p-q)^n\pi_0(0) + (1-p)q\sum_{j=0}^{n-1}(1-p-q)^j \\
&\quad +q - q(1-p-q)^n\pi_0(0) \quad - q^2\sum_{j=0}^{n-1}(1-p-q)^j \\
&= (1-p-q)(1-p-q)^n\pi_0(0) + (1-p-q)q\sum_{j=0}^{n-1}(1-p-q)^j + q \\
&= (1-p-q)^{n+1}\pi_0(0) + q\left(\sum_{j=0}^{n-1}(1-p-q)^{j+1} + 1\right) \\
&= (1-p-q)^{n+1}\pi_0(0) + q\sum_{j=0}^{n}(1-p-q)^j,
\end{aligned}$$

which yields the desired result.

Exercise 9.5.7 Consider a machine that at the start of any particular day is either broken down or in operating condition. Assume that if the machine is broken down at the start of the nth day, the probability that it will be successfully repaired and in operating condition at the start of the $(n+1)^{\text{th}}$ day is p. Assume also that if the machine is in operating condition at the start of the nth day, the probability that it will fail and be broken down at the start of the $(n+1)^{\text{th}}$ day is q. Let $\pi_0(0)$ denote the probability that the machine is broken down initially.

(a) Find the following probabilities

$$\begin{aligned}
&\text{Prob}\{\ X_{n+1} = 1 \mid X_n = 0\ \}, \\
&\text{Prob}\{\ X_{n+1} = 0 \mid X_n = 1\ \}, \\
&\text{Prob}\{\ X_{n+1} = 0 \mid X_n = 0\ \}, \\
&\text{Prob}\{\ X_{n+1} = 1 \mid X_n = 1\ \} \ \text{ and} \\
&\text{Prob}\{\ X_0 = 1\ \}.
\end{aligned}$$

(b) Compute $\text{Prob}\{X_n = 0\}$ and $\text{Prob}\{X_n = 1\}$ in terms of p, q, and $\pi_0(0)$.

(c) Find the steady-state distributions $\lim_{n\to\infty} \text{Prob}\{X_n = 0\}$ and $\lim_{n\to\infty} \text{Prob}\{X_n = 1\}$.

Answer 9.5.7 Let state 0 correspond to the machine's being broken down and let state 1 correspond to the machine being in operating condition. Let X_n be the random variable denoting the state of the machine at day n. We know that $\text{Prob}\{X_0 = 0\} = \pi_0(0)$. Then

(a)

$$\begin{aligned}
\text{Prob}\{\ X_{n+1} = 1 \mid X_n = 0\ \} &= p, \\
\text{Prob}\{\ X_{n+1} = 0 \mid X_n = 1\ \} &= q, \\
\text{Prob}\{\ X_{n+1} = 0 \mid X_n = 0\ \} &= 1 - p, \\
\text{Prob}\{\ X_{n+1} = 1 \mid X_n = 1\ \} &= 1 - q, \\
\text{Prob}\{\ X_0 = 1\ \} &= 1 - \pi_0(0).
\end{aligned}$$

(b)

$$\begin{aligned}
\text{Prob}\{X_{n+1} = 0\} &= \text{Prob}\{X_{n+1} = 0 \mid X_n = 0\}\text{Prob}\{X_n = 0\} + \\
&\quad \text{Prob}\{X_{n+1} = 0 \mid X_n = 1\}\text{Prob}\{X_n = 1\}.
\end{aligned}$$

Let $n = 0$;

$$\begin{aligned}
\text{Prob}\{X_1 = 0\} &= \text{Prob}\{X_1 = 0 \mid X_0 = 0\}\text{Prob}\{X_0 = 0\} + \\
&\quad \text{Prob}\{X_1 = 0 \mid X_0 = 1\}\text{Prob}\{X_0 = 1\} \\
&= (1-p)\pi_0(0) + q[1 - \pi_0(0)] \\
&= (1 - p - q)\pi_0(0) + q.
\end{aligned}$$

Let $n = 1$;

$$\begin{aligned}
\text{Prob}\{X_2 = 0\} &= \text{Prob}\{X_2 = 0 \mid X_1 = 0\}\text{Prob}\{X_1 = 0\} + \\
&\quad \text{Prob}\{X_2 = 0 \mid X_1 = 1\}\text{Prob}\{X_1 = 1\} \\
&= (1-p)[\,(1-p-q)\pi_0(0) + q\,] \\
&\quad +q[\,1 - (1-p-q)\pi_0(0) - q\,] \\
&= (1-p-q)[\,(1-p-q)\pi_0(0) + q\,] + q \\
&= (1-p-q)^2\pi_0(0) + (1-p-q)q + q.
\end{aligned}$$

Repeating n times

$$\begin{aligned}
\text{Prob}\{X_n = 0\} &= (1-p-q)^n\pi_0(0) + q\sum_{j=0}^{n-1}(1-p-q)^j, \\
\text{Prob}\{X_n = 1\} &= 1 - \text{Prob}\{X_n = 0\}.
\end{aligned}$$

(c) Since p and q are probabilities, $0 \le p,\ q \le 1$. Notice that if p and q are both equal to 0 or both equal to 1, the problem is trivial. This only leaves the case when $0 < p + q < 2$.

Observe that since $p + q > 0$, $\sum_{j=0}^{n-1}(1-p-q)^j$ is the sum of a finite geometric progression. Therefore

$$\sum_{j=0}^{n-1}(1-p-q)^j = \frac{1-(1-p-q)^n}{p+q}.$$

$$\begin{aligned}
\text{Prob}\{X_n = 0\} &= (1-p-q)^n\pi_0(0) + \frac{q}{p+q} - q\frac{(1-p-q)^n}{p+q} \\
&= \frac{q}{p+q} + (1-p-q)^n\left[\pi_0(0) - \frac{q}{p+q}\right].
\end{aligned}$$

Also

$$\text{Prob}\{X_n = 1\} = \frac{p}{p+q} + (1-p-q)^n\left[\pi_0(1) - \frac{p}{p+q}\right].$$

Since $\mid 1-p-q \mid < 1$, $\lim_{n\to\infty}(1-p-q)^n = 0$ and

$$\lim_{n\to\infty}\text{Prob}\{X_n = 0\} = \frac{q}{p+q} \quad \text{and} \quad \lim_{n\to\infty}\text{Prob}\{X_n = 1\} = \frac{p}{p+q}.$$

Alternatively, (and much more easily) we can solve

$$(p_0,\ \ p_1)\begin{pmatrix} 1-p & p \\ q & 1-q \end{pmatrix} = (p_0,\ \ p_1)$$

to get the steady-state distribution, i.e.,

$$p_0(1-p) + p_1 q = p_0$$

$$p_0 p + p_1(1-q) = p_1$$

$$p_0 + p_1 = 1.$$

Together these imply that

$$p_0 = \frac{q}{p+q} \quad \text{and} \quad p_1 = \frac{p}{p+q}.$$

Exercise 9.6.1 The time between arrivals at a service center is exponentially distributed with a mean interarrival time of 15 minutes. Each arrival brings one item to be serviced with probability p $(0 < p < 1)$ or two items with probability $1 - p$. Items are serviced individually at a rate of five per hour. Represent this system as a continuous-time Markov chain and give its infinitesimal generator Q.

Answer 9.6.1 We may represent this system as a continuous-time Markov chain by allowing the integers $i = 0, 1, \ldots,$ to represent the number of items in the system. If at any time there is one or more items, then one is being processed and the others are waiting to be processed. In any state $n \geq 0$, an arrival either increases the number to $n+1$ with probability p or increases it to $n+2$ with probability $1-p$. A service completion in state $n > 0$ decreases the number to $n-1$. The infinitesimal generator (with $\lambda = 4$ and $\mu = 5$) is then

$$Q = \begin{pmatrix} -\lambda & p\lambda & (1-p)\lambda & 0 & 0 & 0 & \cdots \\ \mu & -(\lambda+\mu) & p\lambda & (1-p)\lambda & 0 & 0 & \cdots \\ 0 & \mu & -(\lambda+\mu) & p\lambda & (1-p)\lambda & 0 & \cdots \\ 0 & 0 & \mu & -(\lambda+\mu) & p\lambda & (1-p)\lambda & \\ \vdots & \vdots & & \ddots & \ddots & \ddots & \ddots \end{pmatrix}.$$

www.ingramcontent.com/pod-product-compliance
Lightning Source LLC
Chambersburg PA
CBHW080300180526
45167CB00006B/2614

* 9 7 8 1 4 9 9 5 0 8 7 4 1 *